New Perspectives and Approaches in Plant Growth-Promoting Rhizobacteria Research

New Perspectives and Approaches in Plant Growth-Promoting Rhizobacteria Research

Edited by:

P.A.H.M. Bakker, J.M. Raaijmakers, G. Bloemberg, M. Höfte, P. Lemanceau and B.M. Cooke

Reprinted from *European Journal of Plant Pathology,* Volume 119 Issue 3, 2007

Springer

A C.I.P. catalogue record for this book is available from the libray of Congress

ISBN 978-90-481-7722-6
e-ISBN 978-1-4020-6776-1

Published by Springer,
P.O. Box 17, 3300 AA, Dordrecht, The Netherlands

Printed on acid-free paper

Cover photos:
From top to bottom: Confocal laser scanning microscopy analyses of colonies of *Pseudomonas fluorescens* WCS365 marked with green and cyan fluorescent proteins; Model of the transcriptional regulation of ACC deaminase expression in *Pseudomonas putida* UW4; Overview of interactions between biocontrol strains, plants, pathogens, predators, cooperators, and soil; Circular representation of the genome of *Pseudomonas fluorescens* Pf-5

Springeronline.com

Softcover reprint of the hardcover 1st edition 2007

European Journal of Plant Pathology

Volume 119 · Number 3 · November 2007

Special Issue: New perspectives and approaches in Plant Growth-Promoting Rhizobacteria research
Edited by: P.A.H.M. Bakker · J.M. Raaijmakers · G. Bloemberg · M. Höfte · P. Lemanceau · B.M. Cooke

Instructions for Authors for *Eur J Plant Pathol* are available at http://www.springer.com/10658.

Eur J Plant Pathol (2007) 119:241–242
DOI 10.1007/s10658-007-9114-z

FOREWORD

Foreword

Peter A. H. M. Bakker · Jos M. Raaijmakers · Guido V. Bloemberg · Monica Höfte · Philippe Lemanceau · Mike Cooke

Received: 15 February 2007 / Accepted: 15 February 2007 / Published online: 23 March 2007

New perspectives and approaches in plant growth-promoting rhizobacteria research

Plant growth-promoting rhizobacteria (PGPR) are defined as root-colonizing bacteria that exert beneficial effects on plant growth and development. Root colonization comprises the ability of bacteria to establish on or in the plant root, to propagate, survive and disperse along the growing root in the presence of the indigenous microflora. Rhizobacteria are considered as efficient microbial competitors in the root zone. Representatives of many different bacterial genera have been commercialized and/or introduced into soils, onto seeds, roots, tubers or other planting materials to improve crop growth. These bacterial genera include *Acinetobacter, Agrobacterium, Arthrobacter, Azospirillum, Bacillus, Bradyrhizobium, Frankia, Pseudomonas, Rhizobium, Serratia, Thiobacillus* and others. To date, probably the most widely used PGPR in agriculture are *Rhizobium* and *Bradyrhizobium* species for their nitrogen-fixing capacity in roots of Leguminosae. In addition to the promotion of plant growth, PGPR are also employed for controlling plant pathogens, enhancing efficiency of fertilizers, and degrading xenobiotic compounds (rhizoremediation). The application of PGPR is a growing market.

There is an active and growing group of scientists working on fundamental and applied aspects of PGPR. Since the late eighties,

P. A. H. M. Bakker
Faculty of Biology, Section Phytopathology, Utrecht University, PO Box 80084, TB Utrecht 3508, The Netherlands
e-mail: P.A.H.M.Bakker@bio.uu.nl

J. M. Raaijmakers
Laboratory of Phytopathology, Wageningen University, Binnenhaven 5, PD Wageningen, The Netherlands
e-mail: jos.raaijmakers@wur.nl

G. V. Bloemberg
Institute of Biology, Clusius Laboratory, Leiden University, Wassenaarsewegc 64, AL Leiden 2333, The Netherlands
e-mail: g.v.bloemberg@biology.leidenuniv.nl

M. Höfte
Pytopathology Lab, Faculty of Bioscience Engineering, University Ghent, Coupure Links 653, Ghent B-9000, Belgium
e-mail: Monica.Hofte@Ugent.be

P. Lemanceau (✉) · M. Cooke
UMR 1229 Microbiologie du Sol et de l'Environnement, INRA, Université de Bourgogne, CMSE, 17 rue Sully, BP 86510, Dijon Cedex F-21065, France
e-mail: lemancea@dijon.inra.fr

developments of PGPR research have been addressed at International Workshops on PGPR. The first meeting was held in 1987 in Orilla, Ontario, Canada, and since then in Interlaken, Switzerland (1990), Adelaide, Australia (1994), Sapporo, Japan (1997), Cordoba, Argentina (2000) and Calicut, India (2003). In 2006, the 7th workshop was organized in Noordwijkerhout, The Netherlands, where over 130 scientists from 17 countries worldwide participated and presented their results in 49 oral and 69 poster presentations.

Topics addressed during the PGPR workshops include:

- mechanisms of plant growth promotion and disease suppression
- traits involved in root colonization by PGPR
- the role of PGPR in microbial interactions
- the molecular and biochemical basis of disease suppression and root colonization
- the role of PGPR in disease-suppressive soils
- plant responses to PGPR
- discovery of novel PGPR strains and traits
- pathogen responses to PGPR
- risk assessment of PGPR
- production, formulation and delivery strategies of PGPR
- performance of PGPR in greenhouse trials and agricultural fields
- registration and commercialization of PGPR

In addition to these topics the 7th meeting focused on recent developments in genomics, proteomics and metabolomics of PGPR. The abstract book is available at http://www.bio.uu.nl/~fytopath/PDF%20files/abstract%20book%20PGPR%20final.pdf. Last but surely not least, this meeting was dedicated to the great efforts of several PGPR scientists. These are Jim Cook, Geneviève Défago, Ben Lugtenberg and Kees van Loon. In this special issue of the European Journal of Plant Pathology, key contributions are published that give an overview of the work presented at the workshop.

The attendance and excellent contributions by an ever-growing group of young scientists guarantees a healthy future for PGPR research. Our best wishes to David Weller and Joyce Loper who will organize the next workshop in the Pacific Northwest, USA.

Eur J Plant Pathol (2007) 119:243–254
DOI 10.1007/s10658-007-9165-1

REVIEW PAPER

Plant responses to plant growth-promoting rhizobacteria

L. C. van Loon

Received: 4 December 2006 / Accepted: 3 May 2007 / Published online: 5 June 2007

Abstract Non-pathogenic soilborne microorganisms can promote plant growth, as well as suppress diseases. Plant growth promotion is taken to result from improved nutrient acquisition or hormonal stimulation. Disease suppression can occur through microbial antagonism or induction of resistance in the plant. Several rhizobacterial strains have been shown to act as plant growth-promoting bacteria through both stimulation of growth and induced systemic resistance (ISR), but it is not clear in how far both mechanisms are connected. Induced resistance is manifested as a reduction of the number of diseased plants or in disease severity upon subsequent infection by a pathogen. Such reduced disease susceptibility can be local or systemic, result from developmental or environmental factors and depend on multiple mechanisms. The spectrum of diseases to which PGPR-elicited ISR confers enhanced resistance overlaps partly with that of pathogen-induced systemic acquired resistance (SAR). Both ISR and SAR represent a state of enhanced basal resistance of the plant that depends on the signalling compounds jasmonic acid and salicylic acid, respectively, and pathogens are differentially sensitive to the resistances activated by each of these signalling pathways. Root-colonizing *Pseudomonas* bacteria have been shown to alter plant gene expression in roots and leaves to different extents, indicative of recognition of one or more bacterial determinants by specific plant receptors. Conversely, plants can alter root exudation and secrete compounds that interfere with quorum sensing (QS) regulation in the bacteria. Such two-way signalling resembles the interaction of root-nodulating *Rhizobia* with legumes and between mycorrhizal fungi and roots of the majority of plant species. Although ISR-eliciting rhizobacteria can induce typical early defence-related responses in cell suspensions, in plants they do not necessarily activate defence-related gene expression. Instead, they appear to act through priming of effective resistance mechanisms, as reflected by earlier and stronger defence reactions once infection occurs.

Keywords Arabidopsis · Disease suppression · Induced systemic resistance · Plant growth promotion · Signal transduction · Systemic acquired resistance

L. C. van Loon (✉)
Department of Biology, Section Phytopathology, Institute of Environmental Biology, Faculty of Science, Utrecht University, P.O. Box 800.84, 3508 TB Utrecht, The Netherlands
e-mail: l.c.vanloon@uu.nl

Plant growth promotion by rhizobacteria

Plant roots offer a niche for the proliferation of soil bacteria that thrive on root exudates and lysates. Population densities of bacteria in the rhizosphere may be up to 1,00-fold higher than in bulk soil and up to 15% of the root surface may be covered by micro-colonies of a variety of bacterial strains. While these bacteria utilize the nutrients that are released from the host for their growth, they also secrete metabolites into the rhizosphere. Several of these metabolites can act as signalling compounds that are perceived by neighbouring cells within the same micro-colony, by cells of other bacteria that are present in the rhizosphere, or by root cells of the host plant (Van Loon and Bakker 2003; Bais et al. 2004; Gray and Smith 2005; Kiely et al. 2006).

The best-studied example of signal exchange is the *Rhizobium*—legume symbiosis, in which the plant releases flavonoid compounds that act as signals for the bacterium to secrete Nod factors. Nod factors are perceived by plant root hairs and function in a hormone-like fashion to induce root nodules in which the *Rhizobium* bacterium can fix atmospheric nitrogen. The bacterium grows at the expense of carbohydrates from the host, but provides fixed nitrogen for amino acid biosynthesis in return (Brencic and Winans 2005; Gray and Smith 2005). This symbiosis is a prime example of an intimate relationship between a soil bacterium and its host plant, and illustrates the concept behind the term 'plant growth-promoting rhizobacteria' (PGPR): in nitrogen-poor environments the *Rhizobium* bacterium promotes legume plant growth by providing a limiting nutrient.

Growth promotion by soil microorganisms is far from uncommon (Glick et al. 1999; Ryu et al. 2005) and can be considered part of a continuum in which interactions between plants and microorganisms range from deleterious (pathogens) to beneficial (PGPR). In the Netherlands, already 75 years ago observations were made by an assistant of Professor Johanna Westerdijk at the Phytopathological Laboratory 'Willie Commelin Scholten' in Baarn, about recovery from damping-off in turfgrass. The person, by the name of Van Luijk, identified several pathogenic *Pythium* species that were responsible for the disease, but he also observed that grass seeds germinated to a higher percentage in non-sterile than in sterilized soil (Van Luijk 1938). This was the first demonstration in the Netherlands that soil microorganisms can promote plant growth. The reason for this stimulatory effect of the biological agent present in the raw soil became clear only later. It turned out that non-pathogenic *Pythium* spp. were also present, took over and counteracted the actions of the pathogenic *Pythium* spp. and other deleterious soil microorganisms through microbial antagonism. These observations were the beginning of a research programme on antagonism between microorganisms that has been continuing to this day at Utrecht University.

The stimulation of seed germination and the recovery from damping-off of the turfgrass that were caused by the non-pathogenic *Pythium* spp. were apparent as a promotion of growth relative to appropriate control plants. However, in reality they were the result of disease suppression. Many bacteria in soil have similar properties (Compant et al. 2005; Haas and Défago 2005), but in a number of cases rhizobacteria can enhance plant growth in the absence of potentially pathogenic microorganisms, as has been shown in e.g. gnotobiotic systems (Van Loon and Bakker 2003). Over the years, several mechanisms of rhizobacterial growth promotion have been documented (Table 1). The ability to fix atmospheric nitrogen is present in various bacterial species that are either free-living in the soil, or associated with plant roots by growing endophytically (Dobbelaere et al. 2003). Poorly soluble inorganic nutrients that are rate-limiting for growth can be made available through the solubilizing action of bacterial siderophores or the secretion of organic acids (Vessey 2003). The high population densities of bacteria in the rhizosphere stimulate nutrient delivery and uptake by plant roots.

Other mechanisms of growth promotion involve modulation of plant regulatory mechanisms through the production of hormones or other compounds that influence plant development (Frankenberger and Arshad 1995). Many bacterial species are capable of producing auxin and/or ethylene, and synthesis of gibberellins and cytokinins has also been documented. Introduction of the rhizobacterial strain *Pseudomonas fluorescens* WCS417 in autoclaved soil promoted growth of *Arabidopsis* accession Col-0 by 33% (Pieterse and Van Loon 1999). A comparable growth promotion was seen when *Arabidopsis* seedlings were grown under gnotobiotic conditions on vertically oriented agar plates containing half-strength Hoagland nutrient medium. Compared to sterile grown control seedlings, WCS417-treated seedlings showed enhanced shoot and root development, enhanced greening and lateral root formation (S. van der Ent unpublished observation). Whether

Table 1 Mechanisms of plant growth promotion by rhizobacteria

Nitrogen fixation	
Ion uptake	
	Iron, zinc, other essential micronutrients
	Phosphate
Production of plant hormones	
	Auxins, gibberellins, cytokinins, ethylene
Modulation of plant development	
	ACC deaminase
	'Elicitors'

WCS417 produces plant hormones is not known, but promotion of lateral root formation is a typical auxin effect (Tanimoto 2005). Obviously, enhanced lateral root formation increases the capacity to take up nutrients. For *Azospirillum brasilense* it has been shown that auxin is responsible for its growth-promoting action in wheat and pearl millet, as bacterial mutants that had lost 70% of their capacity to produce indole–acetic acid had lost their plant growth-promoting activity (Barbieri and Galli 1993).

Gibberellins and cytokinins both stimulate shoot development. Their effects on root growth are less well documented. Ethylene is usually considered an inhibitor of plant growth, but at low levels can actually promote growth in several plant species, including *Arabidopsis* (Pierik et al. 2006). At moderate levels it inhibits both root and shoot elongation, and at high levels it enhances senescence and organ abscission (Abeles et al. 1992). The direct precursor of ethylene in the plant biosynthetic pathway, 1-aminocyclopropane-1-carboxylic acid (ACC) is exuded from plant roots together with other amino acids. Rhizobacteria that possess the enzyme ACC deaminase can degrade ACC and utilize it as a carbon source. Under such conditions, re-uptake by the roots is prevented and the level of ACC in the roots is reduced. As a consequence, ethylene production by the roots is lowered, relieving inhibition of root growth. Thus, ACC deaminase-containing rhizobacteria can increase root growth by lowering endogenous ACC levels (Glick 2005). However, bacteria lacking ACC deaminase have also been shown to increase plant growth and such observations cannot be explained by known mechanisms. It is presumed that under such conditions bacterial cells possess certain surface components or secrete compounds that act as 'elicitors' of plant growth. Plant roots must be able to perceive and recognize such elicitors in ways similar to the recognition of elicitors from plant pathogens. In fact, plant pathogens might interfere with the action of PGPR by being perceived by similar receptors.

Plant-mediated disease suppression by rhizobacteria

When plants are growing naturally in soils, one cannot distinguish whether an apparent growth promotion is caused by bacterially stimulated plant growth or through suppression of deleterious soil microorganisms. Non-pathogenic rhizobacteria can antagonize pathogens through competition for nutrients, production of antibiotics and secretion of lytic enzymes (Handelsman and Stabb 1996; Van Loon and Bakker 2003). Such activities are particularly important in the rhizosphere where pathogenic fungi are attracted to plant roots. However, rhizobacteria can reduce the activity of pathogenic microorganisms not only through microbial antagonism, but also by activating the plant to better defend itself. This phenomenon, termed 'induced systemic resistance' (ISR) was first described by Van Peer et al. (1991) in carnation that was systemically protected against *Fusarium oxysporum* f.sp. *dianthi* upon treatment with strain WCS417, and by Wei et al. (1991) in cucumber, where six out of 94 rhizobacterial strains protected the leaves against anthracnose caused by *Colletotrichum orbiculare*. Protection as a result of microbial antagonism was excluded because the inducing rhizobacteria and the challenging pathogens were inoculated at, and remained confined to, spatially separated parts on the same plants. Hence, the protective effect was plant-mediated.

ISR confers on the plant an enhanced defensive capacity (Van Loon et al. 1998; Van Loon and Bakker 2005). Upon infection with a challenging pathogen this enhanced defensive capacity is manifested as a reduction in the rate of disease development, resulting in fewer diseased plants or in lesser disease severity. The induced resistance is also evident locally and sometimes does not extend systemically (Van Loon 2000). When only local, it is difficult to prove, because the inducing bacterium and the challenging pathogen are not separated from each other and direct antagonism is difficult to rule out. Only when specific eliciting components of the inducer are active in stimulating resistance in the plant but inactive in antagonizing the pathogen in vitro on different types of media, can locally induced resistance be inferred. Induction of resistance by live organisms always requires proof that the organisms cannot contact each other, a condition that can be met when an inducing rhizobacterium remains confined to the roots and the challenging pathogen colonizes only the leaves. Under such situations the inducing bacterium must trigger the roots to locally produce a signal that moves to the leaves to activate the enhanced defensive capacity systemically. The nature of this mobile signal has so far remained elusive.

Since its discovery, rhizobacteria-mediated ISR has been documented in at least 15 plant species (Van Loon and Bakker 2006). Its induction has been shown to share several characteristics (Table 2A), but its expression can involve different physiological mechanisms (Table 2B). ISR can be induced by various non-pathogenic microorganisms and by some types of stress that activate the same response in the plant. In contrast to *R*-gene-mediated resistance, it is not specific but active against all types of pathogens, as well as against several nematodes and insects. Once induced, plants may remain protected for a considerable part of their lifetime, indicating that when the state of ISR has been triggered in the plant, it is rather stable (Van Loon et al. 1998).

Upon challenge inoculation, ISR is expressed as a result of the altered physiological state of the plant. Expression may take different forms, depending on the activity of the inducing rhizobacterium and the nature of the interaction between the pathogen and the plant (Chester 1933). In fact, 'induced resistance' is an operational term to denote a condition in which a plant becomes less diseased compared to a control plant that was not induced. There are many ways in which developmental and environmental factors can influence plant-pathogen interactions. Damping-off due to infection by *Pythium*, *Fusarium* or *Rhizoctonia* is often confined to the seedling stage. Any condition that results in more rapid plant growth will shorten the vulnerable stage and be apparent as enhanced resistance. Rhizobacteria acting through growth promotion could protect plants through this mechanism. A similar type of ISR could occur in potato where accelerated development leads to enhanced adult plant resistance against late blight caused by *Phytophthora infestans* (Visker et al. 2003).

Some reports on ISR have indicated reduced symptom expression in the absence of a reduction in pathogen proliferation. This tolerance of the plant to the pathogen must have a physiological basis. Examples are the reduced damage of *Pythium ultimum*-infected cucumbers and lesser extent of soft rot of potato infected by *Erwinia carotovora* pv. *carotovora* upon prior treatment of the plants with ACC deaminase-containing rhizobacterial strains. By lowering the level of stress ethylene in the plant due to pathogenic attack, ACC deaminase acted synergistically with other mechanisms of biocontrol in reducing symptom development without having an effect on the population density of the pathogen (Wang et al. 2000).

Reduced disease can also be the outcome of alterations in the microbial populations in the rhizosphere as a result of altered host physiology. Numbers of resistance-inducing bacteria may be changed, or their expression of resistance-inducing traits may be altered (Mark et al. 2005). Plants commonly react to root colonization by rhizobacteria by increasing the release of exudates, and quantity and composition of root exudates vary with plant developmental stage (Phillips et al. 2004). Thus, plant growth promotion could alter root exudation. Moreover, rhizobacteria that act as minor pathogens or are perceived by the plant as a potential threat, are likely to change the rate and composition of exudates, and to increase the release of lysates.

The population densities and the diversity of the root microflora may affect the number and activity of resistance-inducing rhizobacteria. Quorum sensing (QS) within and between bacterial populations is a major regulatory mechanism in bacteria to adjust their metabolism to crowded conditions or other changes in the biotic and abiotic environment (Whitehead et al. 2001). Interference with bacterial QS by host plants has been documented. Plants can produce and secrete various compounds that mimic QS signals of bacteria and, thereby, alter bacterial activities in the rhizosphere (Bauer and Mathesius 2004). The ecological diversity and its consequences for metabolic activity of rhizosphere bacteria are only poorly known at present and deserve further investigation.

Table 2 The nature of systemically induced resistance in plants

(A) *Characteristics of induced systemic resistance*	
The defensive capacity of the plant is enhanced through microbial stimulation or similar stresses	
The enhanced defensive capacity is expressed systemically throughout the plant	
Induced systemic resistance is active against fungi, bacteria, viruses and, sometimes, nematodes and insects	
Once induced, systemic resistance is maintained for prolonged periods	
(B) *Mechanisms of induced systemic resistance*	
Developmental, escape: linked to growth promotion	
Physiological, tolerance: reduced symptom expression	
Environmental: associated with microbial antagonism in the rhizosphere; altered plant-insect interactions	
Biochemical, resistance:	induction of cell wall reinforcement,
	Induction of phytoalexins
	Induction of pathogenesis-related proteins
	'Priming' of defence responses (resistance)

Rhizobacteria can also alter plant secondary metabolism, resulting in changed plant-insect relationships. Root colonization of cucumber by four different PGPR reduced the level of cucurbitacin, which acts as a feeding stimulant to cucumber beetles (Zehnder et al. 1997). Similar effects on insects that can transmit viruses, might reduce virus diseases through induced resistance against the insect vector rather than against the virus itself.

Finally, non-pathogenic rhizobacteria may activate inducible defence mechanisms in the plant in a similar way to pathogenic microorganisms. Such mechanisms can include reinforcement of plant cell walls, production of anti-microbial phytoalexins, synthesis of pathogenesis-related proteins (PRs) (Hammond-Kosack and Jones 1996), as well as an enhanced capacity to express these defence responses upon challenge inoculation with a pathogen, a mechanism known as 'priming' (Conrath et al. 2006). Activation of defence reactions suggests that even a beneficial rhizobacterium may be perceived by the plant as a potential threat, and that such perception involves production of resistance-eliciting compounds that act mechanistically similar to elicitors produced by plant pathogenic fungi and bacteria. Both nitrogen-fixing *Rhizobia* in legume root nodules and vesicular–arbuscular (VA) mycorrhizal fungi in roots have been shown to activate plant host defences when the symbiotic interaction becomes unproductive (Parniske et al. 1991; Hause and Fester 2005). Plants possess sensitive mechanisms to perceive both fungi and bacteria through conserved components that are specific to their kingdoms and act as general elicitors. These are commonly referred to as 'pathogen-associated molecular patterns' (PAMPs) (Nürnberger and Lipke 2005). During compatible plant-pathogen interactions and effective symbioses, the microorganisms actively suppress defensive activities in the host (Da Cunha et al. 2006). The relationship between root-colonizing, resistance-inducing PGPR and their hosts seems substantially less intimate than with either *Rhizobia* or mycorrhizal fungi, but the idea that PGPR may at the same time trigger and suppress defence reactions in the host, deserves consideration.

Expression of systemically induced resistance in the plant

Besides biochemical techniques, such as enzyme activity measurements and protein analysis, the development of molecular-biological techniques has allowed the reaction of plants to rhizobacteria to be determined at the transcriptional level by analyzing differential gene expression by a variety of techniques. Changes in a number of host plants in reaction to several resistance-inducing strains have been documented (Table 3). Many authors report increases in stress-related enzyme activities such as phenylalanine ammonia-lyase, peroxidase, polyphenoloxidase, ß-1,3-glucanase and chitinase, as well as induction of specific PRs in leaves of plants of which the roots were colonized by resistance-inducing PGPR (reviewed in Van Loon and Bakker 2005,

Table 3 Changes in gene expression in bacterized plants

Bacterial strain	Host plant	Systemically in leaves		Locally in roots	
		Up	Down	Up	Down
Various	Various	PRs			
Paenibacillus polymyxa B2	*Arabidopsis*	10			
Pseudomonas thivervalensis MLG45	*Arabidopsis*	42	21	0	9
Pseudomonas fluorescens WCS417	*Arabidopsis*	0	0	39	63
Pseudomonas chlororaphis O6	Cucumber	0			
Pseudomonas fluorescens FPT9601-T5	*Arabidopsis*	95	105		
Bacillus subtilis M4	Cucumber	3.7%	2.5%		
	Tomato	6.2%	4.7%		
Pseudomonas fluorescens C7R12	*M. trunculata*			58	

For references and details, see main text

2006). Park and Kloepper (2000) used *Arabidopsis* transformed with a *PR-1* promoter- ß-glucuronidase (*GUS*) reporter construct and monitored GUS expression in response to nine rhizobacterial strains in plants growing either in vitro on agar plates or in vivo in soil. Almost all strains induced the *PR-1* promoter to varying levels, and expression was correlated roughly with known resistance-inducing properties of these strains. Thus, in these *Arabidopsis* plants, resistance-inducing PGPR induced defence reactions commonly associated with pathogen infection.

Similar results were obtained by Timmusk and Wagner (1999), who concluded that the resistance-inducing strain *Paenibacillus polymyxa* B2 induced mild biotic stress. These authors used gnotobiotically-raised plants on a nutrient medium, and by applying RNA differential display and real-time PCR, found six and an additional four genes, respectively, to be upregulated in response to the PGPR, including *PR-1* and the drought-responsive gene *ERD15*. The significance of the latter observation is difficult to assess. However, it is known that water relations of sterile-grown plants are different from those in the natural environment and the bacteria obviously affected the water potential of the roots.

Using cDNA micro-arrays representing approximately 14,300 genes, Cartieux et al. (2003) monitored gene expression in both leaves and roots of axenic *Arabidopsis* plants infected by resistance-inducing *Pseudomonas thivervalensis* strain MLG45. Plants colonized by this rhizobacterium showed decreased photosynthetic rates and reduced growth, indicating that *P. thivervalensis* acted as a minor pathogen rather than a PGPR. This conclusion was supported by the changes in gene expression observed. In the leaves, genes associated with photosynthesis and chloroplast functioning, as well as several unknowns, were downregulated, whereas genes implicated in stress, wounding, oxidative burst, or disease resistance were upregulated. However, no typical *PR* genes were activated. Surprisingly few changes were noted at the site of bacterial colonization, i.e. the roots. Colonized root systems showed an approximately 50% reduction in primary root length and an increase in lateral root formation, but levels of only nine transcripts were reduced and none was elevated compared to control roots.

These results contrast with those of Verhagen et al. (2004), who determined changes in gene expression of *Arabidopsis* plants grown on rock wool in the presence of strain WCS417, using an *Arabidopsis* GeneChip Microarray representing about 8,000 genes. Locally in the roots, substantial changes were found in the expression of 97 genes, including ones involved in cell rescue and defence, metabolism, transcription, cellular communication and signal transduction, particularly those involved in ethylene signalling. No consistent changes were found in the leaves, indicating that the onset of ISR as a result of root colonization by WCS417 is not associated with detectable changes in gene expression in the leaves. Comparable results were obtained by Kim et al. (2004), who, using subtractive hybridisation, did not detect any changes in leaves of cucumber plants grown in sterilized soilless growing medium from seeds coated with *Pseudomonas chlororaphis* O6, a strain that was effective in inducing systemic resistance against target leaf spot caused by *Corynespora cassiicola*.

Like WCS417, *Pseudomonas fluorescens* FPT9601-T5 was found to trigger ISR in *Arabidopsis* against *Pseudomonas syringae* pv. *tomato*. Using an Affymetrix GeneChip probe array containing approximately 22,800 genes, Wang et al. (2005) detected 95 and 105 genes that were up- and downregulated, respectively, in leaves of soil-grown plants that had been root-dipped in a suspension of the bacteria. Changes in root gene expression were not analysed. Strain FPT9601-T5 was originally identified as an endophytic PGPR in tomato. It also promoted the growth of *Arabidopsis* plants, even though it suppressed growth in early stages of root colonization. Among the upregulated genes were ones involved in metabolism, signal transduction and stress responses, including a number of *PR*-like genes. Noteworthy, putative auxin-regulated genes, suggested to be related to the observed growth promotion, and nodulin-like genes were upregulated, whereas some ethylene-responsive genes were downregulated, indicating that some parts of signalling pathways related to plant defence seemed to be suppressed. These observations point to a similarity in the relationship of endophytic PGPR and *Rhizobia* with their host plants, as *Rhizobia* reduce plant ethylene levels during nodule formation (Ma et al. 2003). In its interaction with *Arabidopsis*, FPT9601-T5 may possess intermediate characteristics between WCS417 and *P. thivervalensis* MLG45. However, PRs were not induced in the leaves of plants of which the roots had been colonized by either of the latter two strains (Wang et al. 2005). These data indicate that *Arabidopsis* reacts quite differently to different PGPR, even though all these bacterial strains are able to trigger ISR in this species. In every case, the number of genes with altered expression was modest, perhaps because of stringent selection criteria employed.

Using cDNA-AFLP, Ongena et al. (2005) estimated that in leaves of cucumber and tomato plants 3.7% and 6.2% of all genes were upregulated and 6.2% and 4.7% were downregulated, respectively, in response to root colonization by *Bacillus subtilis* strain M4. As crop plants are estimated to possess even more genes than the model plant *Arabidopsis* (about 25,000) (Bevan and Walsh 2006), those percentages correspond to a total of a few 1,000 genes, substantially more than the numbers described in the other studies. However, the nature of the genes with altered expression levels was not investigated. Analysis of the reaction of tomato to the ISR-eliciting strain *Serratia liquefaciens* MG1, using a macroarray containing cDNA probes of 70 defence-related and signalling genes, revealed enhanced expression of 12 genes. Seven of those coded for PRs, whereas the others were involved in oxidative stress, ethylene signalling, or metabolism (Shuhegger et al. 2006).

In roots of the legume species *Medicago truncula*, Sanchez et al. (2005) found 58 genes to be upregulated in response to colonization by the growth-promoting strain *Pseudomonas fluorescens* C7R12, a number in line with that found by Verhagen et al. (2004) in *Arabidopsis* roots colonized by strain WCS417. Out of 10 of the C7R12-induced genes, 9 were not upregulated in the *M. truncula dmi3* mutant, which is impaired in the signal transduction pathway of the Nod factor from *Sinorhizobium meliloti* as well as in mycorrhization by the fungus *Glomus mosseae*. Of those 10 genes, *S. meliloti* activated only one and inhibited four others in wild-type *M. truncula*, whereas *G. mosseae* activated all 10. These data indicate that *M. truncula* shares common molecular pathways in the perception of *P. fluorescens* and *G. mosseae*, and to a minor extent *S. meliloti*. Thus, root-colonizing *Pseudomonas* spp. appear to activate signalling pathways in the plant in common with symbiotic mycorrhizal fungi and nitrogen-fixing *Rhizobia*.

Signalling pathways of systemically induced resistance

The activation of certain *PR* genes in some, though not all, plant-PGPR interactions suggests that the systemic resistance that is induced by the rhizobacterium is similar to pathogen-induced systemic acquired resistance (SAR) (e.g. Wang et al. 2005). SAR signalling in the plant is dependent on salicylic acid (SA) and the regulatory protein NPR1, as evidenced by the loss of SAR in transgenic *NahG* plants that are unable to accumulate SA, and in the *npr1* mutant (Sticher et al. 1997). The enhanced defensive capacity characteristic of SAR is always associated with the accumulation of PRs. Notably, PR-1 is commonly taken as a marker that SAR has been induced (Kessmann et al. 1994). In *Arabidopsis*,

SA-dependent SAR is typically associated with the activation of three *PR* genes: *PR-1*, *-2* and *-5*. The pathogen *P. syringae* pv. *tomato* induces SAR, together with a strong activation of these genes, which results in a reduction in both the proliferation of, and symptoms of bacterial speck induced by the same pathogen when induced plants are challenge-inoculated. Plants that were root-inoculated with non-pathogenic WCS417 did not show *PR*-gene expression before challenge inoculation, even though they did express the capacity to reduce proliferation of the pathogen and symptoms of bacterial speck disease to similar extents upon challenge inoculation (Pieterse et al. 1996).

Whereas ISR in the leaves was not associated with detectable changes in gene expression, upon challenge inoculation with *P. syringae* pv. *tomato* 82 genes showed an augmented expression pattern in ISR-expressing leaves. Of these, 16 genes were upregulated and 14 downregulated in induced, but not in non-induced plants (Verhagen et al. 2004), indicating that the expression of these genes was altered only in plants expressing ISR. Thus, not only were several genes primed to respond faster or more strongly upon pathogen attack, but also a substantial number were expressed in an ISR-specific way. Of the primed genes, 70% were dependent on jasmonate (JA) and/or ethylene (ET) signalling, 13% were both JA- and/or ET- and SA-dependent, and 17% were regulated differently. None of the genes were SA-dependent. These results confirm earlier findings that WCS417-elicited ISR in *Arabidopsis* is not associated with activation of the SA signalling pathway but requires responsiveness to JA and ET (Pieterse et al. 1998; Van Wees et al. 1999).

In accordance with the differences in signalling pathways, SAR and ISR were found to differ in their effectiveness against different types of attackers. On *Arabidopsis*, pathogens that are sensitive to both SA- and JA/ET-dependent defences, such as the oomycete *Hyaloperonospora parasitica*, the fungus *Fusarium oxysporum*, and the bacteria *P. syringae* and *Xanthomonas campestris*, were restricted by both SAR and ISR. In contrast, only ISR was active against the necrotrophic fungi *Alternaria brassicicola* and *Botrytis cinerea*, whereas only SAR was effective against *Turnip crinkle virus* (Ton et al. 2002). As to insect attackers, neither SAR nor ISR reduced feeding damage by larvae of the cabbage white butterfly *Pieris rapae*, whereas both were effective against the beet army worm *Spodoptera exigua* (V. R. van Oosten personal communication). Hence, the spectrum of effectiveness of SAR and ISR is only partly overlapping, reflecting the different signalling pathways involved.

Induction of systemically induced resistance in the plant

By using the available SA-non-accumulating *NahG* transformants in *Arabidopsis*, tobacco and tomato, and various plant mutants impaired in JA or ET signalling, the dependence of systemically induced resistance elicited by various PGPR in these three plant species on SA, JA and/or ET has been determined. ISR elicited by almost all strains was found to be SA-independent, also by strains such as *P. fluorescens* CHA0 and *Serratia marcescens* 90-66, that can themselves produce SA as an additional siderophore (reviewed in Van Loon and Bakker 2005). Only the systemic resistance induced by *Pseudomonas aeruginosa* strain 7NSK2 was SA-dependent (De Meyer et al. 1999). For tomato it was established that it is not the SA which is produced by this strain that triggers ISR, but synthesis of the SA-containing siderophore pyochelin and the antibiotic pyocyanin. In combination, pyochelin and pyocyanin induce the formation of oxygen free radicals in the roots, which triggers SA production in the plant and subsequent activation of an SA-dependent enhanced resistance (Audenaert et al. 2002). *Bacillus* spp. activate some of the same signalling pathways as *Pseudomonas* spp., but can also trigger additional pathways that act independently of NPR1 (Kloepper et al. 2004).

PGPR that elicit ISR in one plant species, may not do so in another, again indicating specificity in the interaction between rhizobacteria and plants. Whereas generally rhizobacteria are not dainty in colonizing roots of different plant species, the perception by the plant of bacterial determinants that trigger ISR appears to be quite specific (Bakker et al. 2003; Meziane et al. 2005; Van Loon and Bakker 2005). Apparently, one or more bacterial components need to be recognized by specific plant receptors. Of three strains, *Pseudomonas putida* WCS358 and *P. fluorescens* WCS374 and WCS417, none is active in eliciting ISR in all out of six plant species

(Table 4), even though levels of root colonization are similar. Remarkably, in *Arabidopsis* strain WCS374 was differentially active in eliciting ISR against different pathogens depending on bioassay conditions (M. Djavaheri personal communication), suggesting that the type and effectiveness of the systemic resistance that is induced by this rhizobacterium is variable.

Failure to elicit ISR on certain hosts may be due to the absence of production of inducing components in the rhizosphere, or to an inability of the particular plant species to perceive such compounds. Usually, a minimum population density of 10^5 colony-forming units per gram of root is required for ISR to be triggered (Raaijmakers et al. 1995), suggesting that QS signals may be necessary for the production of the eliciting compounds by the bacteria. Certain host plants have been shown to interfere with bacterial QS in the rhizosphere (Bauer and Mathesius 2004), and this could impede production of elicitors of ISR. Otherwise, root exudates may not provide critical compounds for elicitor production by the bacterium. Lack of perception has been shown to be responsible for the absence of defence-related reactions in *Arabidopsis* accession Ws-0 in reaction to the bacterial PAMP flagellin (Gómez-Gómez and Boller 2002; Zipfel et al. 2004). Flagellin contains a widely conserved 22-amino acid peptide that is recognized by a LRR-receptor kinase in the plasma membrane of *Arabidopsis* accessions such as the commonly used Col-0. The receptor is lacking in accession Ws-0. However, Ws-0 is also impaired in ethylene sensitivity and, therefore, cannot express ISR (Ton et al. 2001).

By bacterial mutant and complementation analysis several bacterial determinants of ISR elicitation in different plant species have been identified (Table 5). *Arabidopsis* appears particularly prone to induction, as it develops ISR after treatment with cell wall preparations consisting mainly of lipopolysaccharide (LPS), pseudobactin siderophores, flagella (Meziane et al. 2005), the antibiotic 2,4-diacetylphloroglucinol (Iavicoli et al. 2003; Weller et al. 2004), and the volatile metabolite 2,3-butanediol (Ryu et al. 2004). Siderophores and antibiotics are produced by the bacteria to compete for iron and to inhibit other strains in the rhizosphere, respectively, and thus have dual functions in microbial antagonism on the one hand, and elicitation of ISR on the other. The siderophore pyochelin and the antibiotic pyocyanin are both required for the induction of systemic resistance in tomato by *P. aeruginosa* 7NSK2 (Audenaert et al. 2002). QS *N*-acylhomoserine lactones were recently shown to act as inducers of systemic resistance in tomato against *Alternaria alternata* (Shuhegger et al. 2006). The compound 2,3-butanediol is produced by *Bacillus* spp. and not only elicits ISR, but is also involved in promoting growth in *Arabidopsis* (Ryu et al. 2003). How 2,3-butanediol exerts its action and how far both mechanisms are connected, is presently unclear.

The LPS of the biocontrol bacterium *Burkholderia cepacia* has been shown to induce an oxidative burst, as well as a rapid influx of Ca^{2+} and extracellular alkalinization of the medium of tobacco suspension cells (Gerber et al. 2004), all three typical early events in the elicitation of defence responses in plant-pathogen interactions (Nürnberger and Scheel 2001; Garcia-Brugger et al. 2006). Indeed, in whole plants of tobacco (Coventry and Dubery 2001) and *Arabidopsis* (Zeidler et al. 2004), the LPS induced substantial amounts of PRs and activation of *PR* genes, respectively, suggesting that the LPS of

Table 4 Differential induction of systemic resistance by *Pseudomonas* spp. strains

Host plant	WCS358	WCS374	WCS417
Arabidopsis	+	–/+	+
Bean	+	nd	+
Carnation	–	nd	+
Radish	–	+	+
Rice	–	+	–
Tomato	+	nd	+

+, Induced; –, Not induced; nd, Not determined

Table 5 Bacterial determinants of induced systemic resistance in different plant species

Lipopolysaccharides: lipid A; O-antigenic sidechain
Siderophores: pseudobactins; pyochelin; (SA)
Flagella
Antibiotics: pyocyanin, 2,4-diacetylphloroglucinol
N-acylhomoserine lactones
Volatile compounds: 2,3-butanediol

Adapted from Van Loon and Bakker (2005)

B. cepacia acted by activating the SA signalling pathway.

Since ISR elicited by almost all rhizobacterial strains in *Arabidopsis*, tobacco and tomato is SA-independent and not associated with significant activation of *PR* genes, it is still an open question in how far their inducing determinants activate early defence reactions. The variety of eliciting compounds precludes recognition by common receptors. Hence, perception by different receptors may trigger different early signalling events, which may or may not quickly converge into a common response. Even for the LPS, both lipid A (Erbs and Newman 2003) and the O-antigenic side-chain (Leeman et al. 1995) have been shown to be each capable of inducing resistance, but of a different type. Thus, preparations of LPS may activate more than a single pathway, contributing to their effectivenesss in a wide array of plant species. The specificity in the reactions of different plant species to individual strains (Table 4) indicates that the reactions of plants to resistance-inducing PGPR must be the outcome of a dynamic interplay between the production and the perception of ISR-eliciting signals. Whereas some PGPR activate defence-related gene expression, others appear to act solely through priming of effective resistance mechanisms, as reflected by earlier and stronger defence reactions once infection occurs.

References

Abeles, F. B., Morgan, P. W., & Salveit, M. E. Jr. (1992). *Ethylene in plant biology* (2nd ed.). San Diego: Academic Press.

Audenaert, K., Pattery, T., Cornelis, P., & Höfte, M. (2002). Induction of systemic resistance to *Botrytis cinerea* in tomato by *Pseudomonas aeruginosa* 7NSK2: Role of salicylic acid, pyochelin, and pyocyanin. *Molecular Plant-Microbe Interactions, 15*, 1147–1156.

Bais, H. P., Park, S.-W., Weir, T. L., Callaway, R. M., & Vivanco, J. M. (2004) How plants communicate using the underground information superhighway. *Trends in Plant Science, 9*, 26–32.

Bakker, P. A. H. M., Ran, L. X., Pieterse, C. M. J., & Van Loon, L. C. (2003). Understanding the involvement of rhizobacteria-mediated induction of systemic resistance in biocontrol of plant diseases. *Canadian Journal of Plant Pathology, 25*, 5–9.

Barbieri, P., & Galli, E. (1993). Effect on wheat root development of inoculation with an *Azospirillum brasilense* mutant with altered indole-3-acetic acid production. *Research in Microbiology, 144*, 69–75.

Bauer, W. D., & Mathesius, U. (2004). Plant responses to bacterial quorum sensing signals. *Current Opinion in Plant Biology, 7*, 429–433.

Bevan, M., & Walsh, S. (2006). The *Arabidopsis* genome: A foundation for plant research. *Genome Research, 15*, 1632–1642.

Brencic, A., & Winans, S. C. (2005). Detection of and response to signals involved in host-microbe interactions by plant-associated bacteria. *Microbiology and Molecular Biology Reviews, 69*, 155–194.

Cartieaux, F., Thibaud, M. C., Zimmerli, L., Lessard, P., Sarrobert, C., David, P., Gerbaud, A., Robaglia, C., Somerville, S., & Nussaume, L. (2003). Transcriptome analysis of *Arabidopsis* colonized by a plant-growth promoting rhizobacterium reveals a general effect on disease resistance. *The Plant Journal, 36*, 177–188.

Chester, K. S. (1933). The problem of acquired physiological immunity in plants (continued). *The Quarterly Review of Biology, 8*, 275–324.

Compant, S., Duffy, B., Nowak, J., Clément, C., & Barka, E. A. (2005). Use of plant growth-promoting bacteria for biocontrol of plant diseases: Principles, mechanisms of action, and future prospects. *Applied and Environmental Microbiology, 71*, 4951–4959.

Conrath, U., Beckers, G. J. M., Flors, V., García-Agustín, P., Jakab, G., Mauch, F., Newman M.-A., Pieterse, C. M. J., Poinssot, B., Pozo, M. J., Pugin, A., Schaffrath, U., Ton, J., Wendehenne, D., Zimmerli, L., & Mauch-Mani, B. (2006). Priming: Getting ready for battle. *Molecular Plant-Microbe Interactions, 19*, 1062–1071.

Coventry, H. S., & Dubery, I. A. (2001). Lipopolysaccharides from *Burkholderia cepacia* contribute to an enhanced defensive capacity and the induction of pathogenesis-related proteins in *Nicotianae tabacum*. *Physiological and Molecular Plant Pathology, 58*, 149–158.

Da Cunha, L., McFall, A. J., & Mackey, D. (2006). Innate immunity in plants: A continuum of layered defenses. *Microbes and Infection, 8*, 1372–1381.

De Meyer, G., Audenaert, K., & Höfte, M. (1999). *Pseudomonas aeruginosa* 7NSK2-induced systemic resistance in tobacco depends on *in planta* salicylic acid accumulation but is not associated with PR1a expression. *European Journal of Plant Pathology, 105*, 513–517.

Dobbelaere, S., Vanderleyden, J., & Okon, Y. (2003). Plant growth-promoting effects of diazotrophs in the rhizosphere. *Critical Reviews in Plant Sciences, 22*, 107–149.

Erbs, G., & Newman, M. A. (2003). The role of lipopolysaccharides in induction of plant defence responses. *Molecular Plant Pathology, 4*, 421–425.

Frankenberger, W. T., & Arshad, M. (1995). *Phytohormones in soils—microbial production and function*. New York: Marcel Dekker.

Garcia-Brugger, A., Lamotte, O., Vandelle, E., Bourque, S., Lecourieux, D., Poinssot, B., Wendehenne, D., & Pugin, A. (2006). Early signaling events induced by elicitors of plant defenses. *Molecular Plant-Microbe Interactions, 19*, 711–724.

Gerber, I. B., Zeidler, D., Durner, J., & Dubery, I. A. (2004). Early perception responses of *Nicotiana tabacum* cells in response to lipopolysaccharides from *Bulkholderia cepacia*. *Planta, 218*, 647–657.

Glick, B. R. (2005). Modulation of plant ethylene levels by the bacterial enzyme ACC deaminase. *FEMS Microbiology Letters, 251*, 1–7.

Glick, B. R., Patten, C. L., Holguin, G., & Penrose, D. M. (1999). *Biochemical and genetic mechanisms used by plant growth promoting bacteria*. London: Imperial College Press.

Gómez-Gómez, L., & Boller, T. (2002). Flagellin perception: A paradigm for innate immunity. *Trends in Plant Science, 7*, 251–256.

Gray, E. J., & Smith, D. L. (2005). Intracellular and extracellular PGPR: Commonalities and distinctions in the plant-bacterium signaling process. *Soil Biology & Biochemistry, 37*, 395–412.

Haas, D., & Défago, G. (2005). Biological control of soil-borne pathogens by fluorescent pseudomonads. *Nature Reviews Microbiology, 3*, 307–319.

Hammond-Kosack, K. E., & Jones, J. D. G. (1996). Resistance gene-dependent plant defense responses. *The Plant Cell, 8*, 1773–1791.

Handelsman, J., & Stabb, E. V. (1996). Biocontrol of soilborne plant pathogens. *The Plant Cell, 8*, 1855–1869.

Hause, B., & Fester, T. (2005). Molecular and cell biology of arbuscular mycorrhizal symbiosis. *Planta, 221*, 184–196.

Iavicoli, A., Boutet, E., Buchala, A., & Métraux, J.-P. (2003). Induced systemic resistance in *Arabidopsis thaliana* in response to root inoculation with *Pseudomonas fluorescens* CHA0. *Molecular Plant-Microbe Interactions, 16*, 851–858.

Kessmann, H., Staub, T., Ligon, J., Oostendorp, M., & Ryals, J. (1994). Activation of systemic acquired disease resistance in plants. *European Journal of Plant Pathology, 100*, 359–369.

Kiely, P. D., Haynes, J. M., Higgins, C. H., Franks, A., Mark, G. L., Morrissey, J. P., & O'Gara, F. (2006). Exploiting new systems-based strategies to elucidate plant-bacterial interactions in the rhizosphere. *Microbial Ecology, 51*, 257–266.

Kim, M. S., Kim, Y. C., & Cho, B. H. (2004). Gene expression analysis in cucumber leaves primed by root colonization with *Pseudomonas chlororaphis* O6 upon challenge-inoculation with *Corynespora cassiicola*. *Plant Biology, 6*, 105–108.

Kloepper, J. W., Ryu, C.-M., & Zhang, S. A. (2004). Induced systemic resistance and promotion of growth by *Bacillus* spp. *Phytopathology, 94*, 1259–1266.

Leeman, M., Van Pelt, J. A., Den Ouden, F. M., Heinsbroek, M., Bakker, P. A. H. M., & Schippers, B. (1995). Induction of systemic resistance against fusarium wilt of radish by lipopolysaccharides of *Pseudomonas fluorescens*. *Phytopathology, 85*, 1021–1027.

Ma, W. B., Guinel, F. C., & Glick, B. R. (2003). *Rhizobium leguminosarum* biovar *viciae* 1-aminocyclopropane-1-carboxylate deaminase promotes nodulation of pea plants. *Applied and Environmental Microbiology, 69*, 4396–4402.

Mark, G. L., Dow, J. M., Kiely, P. D., Higgins, H., Haynes, J., Baysse, C., Abbas, A., Foley, T., Franks, A., Morrissey, J., & O'Gara, F. (2005). Transcriptome profiling of bacterial responses to root exudates identifies genes involved in microbe-plant interactions. *Proceedings of the National Academy of Sciences of the United States of America, 102*, 17454–17459.

Meziane, H., Van der Sluis, I., Van Loon, L. C., Höfte, M., & Bakker, P. A. H. M. (2005). Determinants of *Pseudomonas putida* WCS358 involved in inducing systemic resistance in plants. *Molecular Plant Pathology, 6*, 177–185.

Nürnberger, T., & Lipka, V. (2005). Non-host resistance in plants: New insights into an old phenomenon. *Molecular Plant Pathology, 6*, 335–345.

Nürnberger, T., & Scheel, D. (2001). Signal transmission in the plant immune response. *Trends in Plant Science, 6*, 372–379.

Ongena, M., Duby, F., Jourdan, E., Beaudry, T., Jadin, V., Dommes, J., & Thonart, P. (2005). *Bacillus subtilis* M4 decreases plant susceptibility towards fungal pathogens by increasing host resistance associated with differential gene expression. *Applied Microbiology and Biotechnology, 67*, 692–698.

Park, K. S., & Kloepper, J. W. (2000). Activation of PR-1a promoter by rhizobacteria which induce systemic resistance in tobacco against *Pseudomonas syringae* pv. *tabaci*. *Biological Control, 18*, 2–9.

Parniske, M., Fischer, H.-M., Hennecke, H., & Werner, D. (1991). Accumulation of the phytoalexin glyceollin I in soybean nodules infected by a *Bradyrhizobium japonicum nifA* mutant. *Zeitschrift für Naturforschung, 46*, 318–320.

Phillips, D. A., Fox, T. C., King, M. D., Bhuvaneswari, T. V., & Teuber, L. R. (2004). Microbial products trigger amino acid exudation from plant roots. *Plant Physiology, 136*, 2887–2894.

Pierik, R., Tholen, D., Poorter, H., Visser, E. J. W., & Voesenek, L. A. C. J. (2006). The Janus face of ethylene: Growth inhibition and stimulation. *Trends in Plant Science, 11*, 176–183.

Pieterse, C. M. J., & Van Loon, L. C. (1999). Salicylic acid-independent plant defence pathways. *Trends in Plant Science, 4*, 52–58.

Pieterse, C. M. J., Van Wees, S. C. M., Hoffland, E., Van Pelt, J. A., & Van Loon, L. C. (1996). Systemic resistance in *Arabidopsis* induced by biocontrol bacteria is independent of salicylic acid accumulation and pathogenesis-related gene expression. *The Plant Cell, 8*, 1225–1237.

Pieterse, C. M. J., Van Wees, S. C. M., Van Pelt, J. A., Knoester, M., Laan, R., Gerrits, H., Weisbeek, P. J., & Van Loon, L. C. (1998). A novel signaling pathway controlling induced systemic resistance in *Arabidopsis*. *The Plant Cell, 10*, 1571–1580.

Raaijmakers, J. M., Leeman, M., Van Oorschot, M. P. M., Van der Sluis, I., Schippers, B., & Bakker, P. A. H. M. (1995). Dose-response relationships in biological control of fusarium wilt of radish by *Pseudomonas* spp. *Phytopathology, 85*, 1075–1081.

Ryu, C.-M., Farag, M. A., Hu, C. H., Reddy, M. S., Wei, H. X., Paré, P. W., & Kloepper, J. W. (2003). Bacterial volatiles promote growth in *Arabidopsis*. *Proceedings of the National Academy of Sciences of the United States of America, 100*, 4927–4932.

Ryu, C.-M., Farag, M. A., Hu, C. H., Reddy, M. S., Kloepper, J. W., & Paré, P. W. (2004). Bacterial volatiles induce systemic resistance in *Arabidopsis*. *Plant Physiology, 134*, 1017–1026.

Ryu, C.-M., Hu, C.-H., Locy, R. D., & Kloepper, J. W. (2005). Study of mechanisms for plant growth promotion elicited by rhizobacteria in *Arabidopsis thaliana*. *Plant and Soil, 268*, 285–292.

Sanchez, L., Weidmann, S., Arnould, C., Bernard, A. R., Gianinazzi, S., & Gianinazzi-Pearson, V. (2005). *Pseudomonas fluorescens* and *Glomus mosseae* trigger *DMI3*-dependent activation of genes related to a signal transduction pathway in roots of *Medicago truncula*ta. *Plant Physiology, 139*, 1065–1077.

Shuhegger, R., Ihring, A., Gantner, S., Bahnweg, G., Knappe, C., Vogg, G., Hutzler, P., Schmid, M., Van Breusegem, F., Eberl, L., Hartmann, A., & Langebartels, C. (2006). Induction of systemic resistance in tomato by *N*-acyl-L-homoserine lactone-producing rhizosphere bacteria. *Plant, Cell and Environment, 29*, 909–918.

Sticher, L., Mauch-Mani, B., & Métraux, J.-P. (1997). Systemic acquired resistance. *Annual Review of Phytopathology, 35*, 235–270.

Tanimoto, E. (2005). Regulation of root growth by plant hormones: Roles for auxin and gibberellin. *Critical Reviews in Plant Sciences, 24*, 249–265.

Timmusk, S., & Wagner, E. G. H. (1999). The plant-growth-promoting rhizobacterium *Paenibacillus polymyxa* induces changes in *Arabidopsis thaliana* gene expression: A possible connection between biotic and abiotic stress responses. *Molecular Plant-Microbe Interactions, 12*, 951–959.

Ton, J., Davison, S., Van Wees, S. C. M., Van Loon, L. C., & Pieterse, C. M. J. (2001). The *Arabidopsis ISR1* locus controlling rhizobacteria-mediated induced systemic resistance is involved in ethylene signaling. *Plant Physiology, 125*, 652–661.

Ton, J., Van Pelt, J. A., Van Loon, L. C., & Pieterse, C. M. J. (2002). Differential effectiveness of salicylate-dependent and jasmonate/ethylene-dependent induced resistance in *Arabidopsis*. *Molecular Plant-Microbe Interactions, 15*, 27–34.

Van Loon, L. C. (2000). Systemic induced resistance. In A. J. Slusarenko, R. S. S. Fraser & L. C. Van Loon (Eds.), *Mechanisms of resistance to plant diseases* (pp. 521–574). Dordrecht, The Netherlands: Kluwer Academic Publishers.

Van Loon, L. C., & Bakker, P. A. H. M. (2003). Signalling in rhizobacteria-plant interactions. In H. De Kroon & E. J. W. Visser (Eds.), *Root ecology, (Ecological studies* Vol. 168, pp. 297–330). Berlin, Heidelberg: Springer-Verlag.

Van Loon, L. C., & Bakker, P. A. H. M. (2005). Induced systemic resistance as a mechanism of disease suppression by rhizobacteria. In Z. A. Siddiqui (Ed.), *PGPR: Biocontrol and biofertilization* (pp. 39–66). Dordrecht, The Netherlands: Springer Science.

Van Loon, L. C., & Bakker, P. A. H. M. (2006). Root-associated bacteria inducing systemic resistance. In S. S. Gnanamanickam (Ed.), *Plant-associated bacteria* (pp. 269–316). Dordrecht, The Netherlands: Springer.

Van Loon, L. C., Bakker, P. A. H. M., & Pieterse, C. M. J. (1998). Systemic resistance induced by rhizosphere bacteria. *Annual Review of Phytopathology, 36*, 453–483.

Van Luijk, A. (1938). Antagonism between various microorganisms and different species of the genus *Pythium*, parasitizing upon grasses and lucerne. *Mededelingen van het Phytopathologisch Laboratorium ''Willie Commelin Scholten''- Baarn, 14*, 43–83.

Van Peer, R., Niemann, G. J., & Schippers, B. (1991). Induced resistance and phytoalexin accumulation in biological control of fusarium wilt of carnation by *Pseudomonas* sp. strain WCS417r. *Phytopathology, 81*, 728–734.

Van Wees, S. C. M., Luijendijk, M., Smoorenburg, I., Van Loon, L. C., & Pieterse, C. M. J. (1999). Rhizobacteria-mediated induced systemic resistance (ISR) in *Arabidopsis* is not associated with a direct effect on expression of known defense-related genes but stimulates the expression of the jasmonate-inducible gene *Atvsp* upon challenge. *Plant Molecular Biology, 41*, 537–549.

Verhagen, B. W. M., Glazebrook, J., Zhu, T., Chang, H. S., Van Loon, L. C., & Pieterse, C. M. J. (2004). The transcriptome of rhizobacteria-induced systemic resistance in *Arabidopsis*. *Molecular Plant-Microbe Interactions, 17*, 895–908.

Vessey, J. K. (2003). Plant growth promoting rhizobacteria as biofertilizers. *Plant and Soil, 255*, 571–586.

Visker, M. H. P. W., Keizer, L. C. P., Budding, D. J., Van Loon, L. C., Colon, L. T., & Struik, P. C. (2003). Leaf position prevails over plant age in reflecting resistance to late blight in potato. *Phytopathology, 93*, 666–674.

Wang, C., Knill, E., Glick, B. R., & Défago, G. (2000). Effect of transferring 1-aminocyclopropane-1-carboxylic acid (ACC) deaminase genes into *Pseudomonas fluorescens* strain CHA0 and its *gacA* derivative CHA96 on their growth-promoting and disease-suppressive capacities. *Canadian Journal of Microbiology, 46*, 898–907.

Wang, Y., Ohara, Y., Nakayashiki, H., Tosa, Y., & Mayama, S. (2005). Microarray analysis of the gene expression profile induced by the endophytic plant growth-promoting rhizobacteria, *Pseudomonas fluorescens* FPT9601-T5 in *Arabidopsis*. *Molecular Plant-Microbe Interactions, 18*, 385–396.

Wei, G., Kloepper, J. W., & Tuzun, S. (1991). Induction of systemic resistance of cucumber to *Colletotrichum orbiculare* by select strains of plant growth-promoting rhizobacteria. *Phytopathology, 81*, 1508–1512.

Weller, D. M., Van Pelt, J. A., Mavrodi, D. V., Pieterse, C. M. J., Bakker, P. A. H. M., & Van Loon, L. C. (2004). Induced systemic resistance (ISR) in *Arabidopsis* against *Pseudomonas syringae* pv. *tomato* by 2,4-diacetylphloroglucinol (DAPG)-producing *Pseudomonas fluorescens*. *Phytopathology, 94*, S108.

Whitehead, N. A., Barnard, A. M., Slater, H., Simpson, N. J., & Salmond, G. P. (2001). Quorum-sensing in Gram-negative bacteria. *FEMS Microbiology Reviews, 25*, 365–404.

Zehnder, G., Kloepper, J., Yao, C. B., & Wei, G. (1997). Induction of systemic resistance in cucumber against cucumber beetles (Coleoptera: *Chrysomelidae*) by plant growth-promoting rhizobacteria. *Journal of Economic Entomology, 90*, 391–396.

Zeidler, D., Zähringer, U, Gerber, I., Dubery, I., Hartung, T., Bors, W., Hutzler, P., & Durner, J. (2004). Innate immunity in *Arabidopsis thaliana*: Lipopolysaccharides activate nitric oxide synthase (NOS) and induce defense genes. *Proceedings of the National Academy of Sciences of the United States of America, 101*, 15811–15816.

Zipfel, C., Robatzek, S., Navarro, L., Oakeley, E. J., Jones, J. D. G., Felix, G., & Boller, T. (2004). Bacterial disease resistance in *Arabidopsis* through flagellin perception. *Nature, 428*, 764–767.

Eur J Plant Pathol (2007) 119:255–264
DOI 10.1007/s10658-007-9201-1

FULL RESEARCH PAPER

Management of resident plant growth-promoting rhizobacteria with the cropping system: a review of experience in the US Pacific Northwest

R. James Cook

Received: 4 December 2006 / Accepted: 26 July 2007 / Published online: 6 September 2007

Abstract In view of the inconsistent performance of single or mixtures of plant growth-promoting rhizobacteria (PGPR) strains formulated for commercial use, and the high cost of regulatory approval for either a proprietary strain intended for disease control or a crop plant transformed to express a disease-suppressive or other growth-promoting PGPR trait, management of resident PGPR with the cropping system remains the most practical and affordable strategy available for use of these beneficial rhizosphere microorganisms in agriculture. A cropping system is defined as the integration of management (agricultural) practices and plant genotypes (species and varieties) to produce crops for particular end-uses and environmental benefits. The build-up in response to monoculture cereals of specific genotypes of *Pseudomonas fluorescens* with ability to inhibit *Gaeumannomyces graminis* var. *tritici* by production of 2,4-diacetylphoroglucinol (DAPG), accounting for take-all decline in the US Pacific Northwest, illustrates what is possible but apparently not unique globally. Other crops or cropping systems enrich for populations of the same or other genotypes of DAPG-producing *P. fluorescens* or, possibly and logically, genotypes with ability to produce one or more of the five other antibiotic or antibiotic-like substances inhibitory to other soilborne plant pathogens. In the U.S Pacific Northwest, maintenance of threshold populations of resident PGPR inhibitory to *G. graminis* var. *tritici* is the centerpiece of an integrated system used by growers to augment take-all decline while also limiting damage caused by pythium and rhizoctonia root rot and fusarium root and crown rot in the direct-seed (no-till) cereal-intensive cropping systems while growing varieties of these cereals (winter and spring wheat, barley and triticale) fully susceptible to all four root diseases.

Keywords Take-all · Take-all decline · *Pseudomonas fluorescens* · Pythium root rot · Rhizoctonia root rot · Fusarium root and crown rot · Direct-seed · Crop rotation · Crop monoculture

R. J. Cook (✉)
Department of Plant Pathology,
Washington State University,
Pullman, WA 99164, USA
e-mail: rjcook@wsu.edu

Introduction

There are basically three strategies by which the biological control and other crop-production benefits of plant growth-promoting rhizobacteria (PGPR) can be exploited: (1) through mass production, formulation and timely introduction of select strains of PGPR, usually with the seeds or other planting material at the time of planting; (2) through transgenic expression in the crop plant itself of one or more specific traits identified with a biological control/growth promotion phenotype; and (3) management of their resident populations and beneficial activities with the cropping system. A *cropping system* is defined as the integration of management (agricultural) practices and plant genotypes (species and varieties) to produce crops for particular end-uses and environmental benefits (Cook 2006).

Initial optimism based on results of pot tests and small-scale field trials that the disease-suppression and other agronomic benefits of PGPR might be attained commercially through use of a single proprietary strain or defined mixture of strains (Cook and Baker 1983) has long since been countered by the realities of their notoriously inconsistent performance from site to site and year to year (Bakker 1989; Cook 1993; Weller 1988). To claim disease control for a proprietary strain intended for commercial use, most countries require that the strain as formulated be registered by the regulatory authority as a microbial pesticide. The cost for attaining and submitting the data necessary to register and label a strain cannot be justified if the performance is inconsistent, especially if, in addition, the strain has activity only against the pathogen or pathogen complex for which it was selected and on the host from which it was selected (Mathre et al. 1999). Activity of a given strain may be broadened through genetic transformation (Blouin Bankhead et al. 2004; Fenton et al. 1992) but this will only further complicate and extend the regulatory approval process.

The use of PGPR as a genetic resource has potential to serve as a source of resistance to soilborne plant pathogens, and should be pursued in view of the dearth of useful genes for resistance to these pathogens within the gene pools of crop plants. Several studies have shown the potential for the *EnCh42* gene from *Trichoderma harzianum* for production of endochitinase as a source of resistance to *Rhizoctonia solani* (Harman et al. 2004). However, unlike the use of *EnCh42*, or the *Bt* gene from *Bacillus thuringiensis* for resistance to certain insect pests, transgenic expression of a functional PGPR trait will likely require the considerably more difficult transfer of a complex multi-gene pathway (Bangera and Thomashow 1999; Mavrodi et al. 1998) and possibly quantitative trait, preferably combined with a tissue- (root-) specific promoter. Moreover, the same or even more comprehensive regulatory requirements must be met for use of a PGRP trait expressed transgenically in the host plant than required for use of the donor PGPR strain itself.

Studies of the ecology of PGPR populations would suggest that resident PGPR populations are likely to respond simply to the agricultural practice of growing crops, whether recognized or unrecognized by science, and may with more knowledge of these interactions be further enhanced by deliberate use of the cropping system with no requirement for regulatory oversight. Indeed, the evidence, albeit empirical, would suggest further that the modern-day success of intensive cropping, including some crop monocultures or near-monocultures, may well be the result of protection from root diseases afforded by naturally occurring PGPR (Cook 2006, 2007). Take-all of wheat caused by *Gaeumannomyces graminis* var. *tritici* and its decline as the soil becomes suppressive in response to continuous wheat or barley monoculture has become a model system for understanding the effects of crop plants and more particularly the cropping system on the populations and activity of PGPR strains of *Pseudomonas fluorescens* (Weller et al. 2002, 2007). Cook et al. (1995) suggested further that, since wheat and related grasses evolved as a monoculture or near-monoculture, natural suppression by PGPR responsible for take-all decline would have been redundant with and therefore could account for the lack of useful genes for resistance to the take-all pathogen. This paper briefly reviews the current state of knowledge of the role of PGPR in take-all decline, their possible role in the success of intensive cropping more generally, and their exploitation as a component of an integrated approach to management of root diseases of wheat and barley in cereal-intensive direct seed (no-till) cropping systems in the US Pacific Northwest.

Role of PGPR in take-all decline

The phenomenon whereby take-all caused by *G. graminis* var. *tritici* increases during the initial years of continuous wheat monoculture but then declines and grain yields increase as the soil becomes suppressive to the pathogen has been well documented since first demonstrated experimentally by Gerlach (1968) in The Netherlands. Research by Shipton et al. (1973) with monoculture wheat-field soils in the Washington State confirmed the findings of Gerlach (1968), including that the factor responsible for pathogen suppression is sensitive to 55°C moist-heat treatment of the soil for 30 min. This work showed further that a take-all suppressive agent present specifically in soil from a field in continuous wheat monoculture was transferable to another field by movement of the soil and that it could multiply (reviewed in Cook 2007). Cook and Rovira (1976) concluded that suppression of the pathogen occurred in the rhizosphere rather than in the bulk soil and suggested that fluorescent *Pseudomonas* species were somehow involved. After some 30 years of research on take-all decline, the evidence now points clearly to a major if not exclusive role of a select few genotypes of *P. fluorescens* with ability to produce the antibiotic 2,4-diacetylphlorglucinol (DAPG) inhibitory to *G. graminis* var. *tritici* (reviewed in Weller et al. 2002, 2007). The threshold population of DAPG-producing psuedomonads required for disease suppression is log 5 colony forming units (CFU)/g root. Raaijmakers et al. (1999) showed further that the amount of DAPG produced in the rhizosphere of wheat is a constant 0.62 ng/log 5 CFU when populations of a wheat-adapted DAPG-producing strain ranged between log 6 and log 7 CFU/g root.

Ability to produce DAPG is a highly conserved trait within the populations of *P. fluorescens* worldwide (Keel et al. 1996). However, based on polymorphisms in the *phlD* gene contained within the five-gene DAPG-biosynthesis operon (Mavrodi et al. 2001), together with DNA fingerprinting, it is possible to detect, quantify, and characterize distinct genotypes of DAPG-producing *P. fluorescens* (McSpadden Gardener et al. 2000, 2001) in the rhizosphere of economically important crops. To date, at least 22 distinct genotypes of DAPG-producing *P. fluorescens* have been identified among the thousands of isolates obtained from rhizospheres (Landa et al. 2005). Among these, the D genotype has been the dominant strain in take-all decline fields in Washington State (Weller et al. 2002, 2007), whereas genotypes F and M were dominant in take-all decline fields in The Netherlands (Souza et al. 2003). Ability to produce DAPG is only one aspect of the effectiveness of a PGPR strain in suppression of take-all. The other aspect is ability to colonize the roots of wheat and sustain a take-all-suppressive threshold population. Strain Q8r1-96, a D genotype, exemplifies both traits; in addition to its ability to produce DAPG, this strain is a highly aggressive colonist of the wheat rhizosphere and was shown to sustain a take-all-suppressive population in competition with other potential rhizosphere colonists in natural soil over successive cycles of growing wheat (Raaijmakers and Weller 2001).

Possible role of PGPR in the success of intensive cropping more generally

Like any new area of science, the fundamental revelations and practical implications of research on take-all decline has been done in concert with and inspired by the discoveries made through research on other examples of plant growth promotion and pathogen suppression by fluorescent *Pseudomonas* species (Schippers et al. 1987). Of particular relevance is work in Switzerland led by G. Dèfago demonstrating a primary role of fluorescent pseudomonads in the suppression of black root rot of tobacco caused by *Thielaviopsis basicola* in a soil cropped 24 years to monoculture tobacco (Stutz et al. 1986). Initial findings with one particular strain isolated from the rhizosphere of tobacco grown in this soil, now the well-known model strain *P. fluorescens* strain CHAO, indicated that the suppression of black root rot with monoculture of the host crop involved a combination of iron starvation by production of siderophores and inhibition of the pathogen by both hydrogen cyanide and antibiotics (Ahl et al. 1986). Subsequent work showed that *P. fluorescens* strain CHAO produces DAPG (Keel et al. 1992), and that up to 23% of 1,100 pseudomonads isolated from the rhizosphere of tobacco grown in the 24-year tobacco-monoculture soil produced DAPG (Keel et al. 1996). *Pseudomonas fluorescens* CHAO is member of the A genotype of DAPG-producing *P. fluorescens.*

Another example is the finding in a plot on the Washington State University, Northwest Research and Extension Centre at Mount Vernon, Washington, that DAPG-producers exceeded log 5.0 CFU/g root in the rhizosphere where peas had been grown in monoculture for the past 30 years and the soil was suppressive to fusarium wilt caused by *Fusarium oxysporum* f. sp. *pisi* (Landa et al. 2002). Six distinct DAPG-producing genotypes were identified in the rhizosphere of pea growing in this soil, with the D and P genotypes being dominant. The D and P genotypes recovered from the pea rhizosphere were shown to colonized the rhizospheres of wheat and pea, respectively, better than the other four DAPG-producing genotypes, namely A, L, O, and Q, also isolated from this pea-monoculture plot.

Further evidence that the different DAPG-producing genotypes may be crop-specific was recently reported by Landa et al. (2005) following their analysis of genotypes present in plots on the campus of North Dakota State University, Fargo, North Dakota, where wheat and flax were grown as monocultures in side-by-side plots for more than 100 years. Populations of DAPG-producing pseudomonads exceeded log 5.0 CFU/g root in the rhizospheres of both wheat and flax grown in the soils. However, roughly 80% of the genotypes in the rhizosphere were of equal frequencies of genotypes F and J in soil where flax had been grown in monoculture, and 77% were genotype D in soil where wheat had been grown in monoculture. The genotype(s) in soil from the plot that had been in crop rotation (i.e. bean, corn, oat, soybean, sugar beet, sunflower etc, or left fallow) for over a century could not be assessed because the population of DAPG-producing PGPR were below the level of detection (10^4 CFU/g root) on roots in this soil.

McSpadden Gardner et al. (2005) found the D genotype in the rhizospheres of both maize and soybeans in Ohio State in all 15 counties sampled over a 3-year period, exceeding log 3.4 CFU/g root on 77, 84 and 81% of maize plants sampled in years 2001, 2002, and 2003, respectively, and 78, 67, and 52% of soybean plants sampled during those 3 years, respectively. Picard et al. (2000) similarly documented high populations of DAPG-producing genotypes of *P. fluoresens* in the rhizosphere of maize in Spain. About half of the 120 million hectares of crops planted each year in the USA are planted to maize and soybeans (about 32 million hectares each). The maize-soybean rotation is the dominant cropping system, but at least 2 million hectares of soybeans are grown in a wheat-soybean double crop system and an equally large area is planted each year to continuous maize. Plant breeding can account, in part, for the continued increases in average annual yield of both crops, including for resistance to the soybean cyst nematode and phytophthora root rot of soybean. However, other than selecting the highest yielding maize inbreds and hybrids under field conditions, there is no deliberate effort by the seed companies to breed for resistance to root diseases of maize. While it can only be speculative at this stage, the pathogen-suppressive benefits of DAGP-producing PGPR enriched, respectively, by the corn–soybean, corn–corn or soybean–wheat double cropping systems and possibly even the PGPR-enriching effects of the corn hybrids and soybean varieties themselves cannot be overlooked and deserve more study.

Crop monoculture has also been shown to convert soil from suppressive to conducive through shifts in populations of rhizobacteria. In Washington State, soils not previously planted to apples are naturally suppressive to a fungal/oomycete complex of apple root pathogens responsible for the apple replant problem but become conducive to these pathogens after apples are grown in the soil (Mazzola 1998). Mazzola (1999) found that, concurrent with enrichment in the inoculum density of root pathogens in response to roots of apple growing in the soil, the populations of microorganisms antagonistic to these pathogens were replaced by pseudomonads not inhibitory to these pathogens, thereby helping to account for the poor growth of apples replanted in this same soil. However, by planting wheat in old apple-orchard soil prior to replanting apples, the population of rhizobacteria returned to predominantly inhibitory pseudomonads and the disease-suppressive state of the soil was restored (Mazzola and Gu 2000).

The diversity of genera, species, subspecies and genotypes of PGPR functioning as resident antagonists of pathogens in soil may well be an order of magnitude greater than the 'tip of the iceberg' revealed to date for the DAPG-producing genotypes of *P. fluorescens*. Even among the diversity of fluorescent pseudomonads, we might assume a similar hierarchy of subspecies and genotypes based on ability to produce one or more of the five other antimicrobial substances produced by strains of this diverse genus (Hass and Défago 2005) and no less conserved globally than ability to produce DAPG. In the same way that the different genotypes of DAPG-producers appear to be adapted to different crops, soils and possibly climates, and suppressive to specific diseases caused by soilborne plant pathogens, it is not a stretch to speculate that other antibiotic-producing genotypes contribute to the suppression of these same or different diseases in agricultural and natural environments. The door to study of the fluorescent pseudomonads was opened with the development by Sands and Rovira (1970) of a selective medium for their isolation and enumeration. The door to the study of other groups, including the unculturables, will depend on the newer methods of isolation and characterization of DNA recovered directly from soil and the rhizosphere of crop plants. This is an exciting yet complex area of rhizosphere microbiology, but one well worth pursuing if science is to more fully understand and exploit the pathogen-suppressive soils as a component of sustainable agriculture.

Exploitation of take-all suppression combined with agricultural practices to manage wheat and barley root diseases in cereal-intensive direct-seed cropping systems

The U.S intermountain Pacific Northwest represents nearly 5 million hectares of farmland in Idaho, Oregon and Washington. With its Mediterranean climate of cool wet winters and dry summers, the region is ideal for but also largely limited to production of cool-season crops, especially wheat and barley. Similar to northern Europe, the yield potential under rainfed conditions in the Palouse region of southeast Washington and adjacent northern Idaho with its deep loess soils is in the range of 8–10 t ha^{-1}, being limited primarily by annual precipitation (Cook and Veseth 1991). Travelling west from the Washington–Idaho border towards the Cascade mountains, annual precipitation drops approximately 1 cm every 10 km and the environments under which wheat and barley are grown change accordingly from subhumid to semi-arid and then arid-irrigated. Yields under irrigation have been documented as high as 14 t ha^{-1}.

Not surprisingly, the conditions ideal for cool-season cereals are also ideal for the diseases of these cereals. Stripe (yellow) rust, pseudocercosporella eyespot, cephalosporium stripe and a snow-mold complex caused by several soil- and crop-residue-borne pathogens have come under control over the past 40 years largely through the use of resistant or tolerant varieties and crop rotation. In contrast, and typical of root diseases more generally, progress towards control of cereal root diseases has depended almost entirely on refinements in agricultural practices, with little if any help from the breeding programmes. Adding to the challenge, decomposition of pathogen-infested roots and stem bases of the host in the top few centimeters of soil (where most concentrated) occurs relatively slowly; when wet enough for microbial breakdown, e.g. late fall to spring, soil temperatures are typically low and when warm enough, e.g., summer and early fall, the soils are dry. The result is that a 1-year break from wheat or barley is not sufficient to control these pathogens by crop rotation. Lengthening the rotation depends on planting cool-season broadleaf crops such as pulses and brassicas, all of which are minor crops that have limited markets, or the production economics is not competitive with production of the same crops in Canada. Adding further to the challenge, growers are moving towards wider adoption of direct seeding (no-till) to lower the cost of fuel, machinery, and time required to produce crops while reversing a 100-year trend of decline in soil organic matter (Rassmusen and Rhode 1989) and preventing soil erosion. The same agricultural practices intended to delay the breakdown of crop residue further increase the potential for crop damage from residue-borne pathogens such as *G. graminis* var. *tritici* (reviewed in Cook 2006). The result is that, while the yield potential for wheat and barley under rainfed conditions is typically higher with direct-seeding compared to the use of conventional tillage, owing to more water captured and kept in the soil, actual yields may be lower due to the increased pressure from root diseases.

Cereal-intensive cropping systems in the Inland PNW include the use of different market classes of wheat (either high-grain-protein bread-type or low-grain-protein pastry type wheat of either a winter or spring growth habit), barley (malting and feed types, two-row and six-row, and either a winter or spring growth habit), and triticale (either winter or spring growth habit). Typical cereal-intensive direct-seed cropping systems vary within the region, depending on the annual precipitation and include: a 3-year winter wheat/spring wheat/spring barley, spring wheat, or chemical-fallow sequence; 2-year winter wheat/spring cereal or chemical-fallow sequence; and different combinations of continuous spring barley, spring wheat and/or spring triticale. Continuous direct-seeded winter cereals leads quickly to unmanageable winter annual grasses such as *Bromus tectorum* and is therefore avoided.

The important root diseases of cereals in these systems include, in addition to take-all, pythium root rot caused by several *Pythium* species; rhizoctonia root rot caused by *R. solani* AG8, *R. oryzae* and possibly one or more other *Rhizoctonia* species; and fusarium root and crown rot caused by *F. pseudograminearum* and *F. culmorum* (Paulitz et al. 2002). Recent findings by Smiley et al. (2004) reveal that plant parasitic nematodes, including *Heterodera avenae, Pratylenchus thornei,* and *P. neglectus,* are also part of the root-pathogen complex limiting yields of dryland winter and spring wheat in the semiarid PNW.

Take-all and pythium and rhizoctonia root rots are favoured by the cool moist soil conditions of late fall and early spring, especially soils covered with residue of the previous crop (typical of direct seed systems) compared to the bare-black soil surface achieved when the crop residue is buried or burned. Indeed, because soil covered with crop residue stays wet longer into the growing season, direct seeding has extended the area favourable to take-all well into the lower precipitation part of the region and beyond its normal range. The occurrence of fusarium root and crown rot requires a period of predisposing water stress on the crop during the reproductive stages of plant development and therefore historically has been most important in the semi-arid parts of the region (Cook 1980), although both take-all and fusarium root and crown rot caused by *F. pseudograminearum* can develop on different plants in the same field in direct-seeded spring wheat in the subhumid parts of the region.

Dependence on spontaneous take-all decline is fundamentally the most important component of the integrated system used to manage root diseases in the direct-seed cereal-intensive cropping systems used in the intermountain Pacific Northwest. Take-all left uncontrolled will dominate the root system, especially of winter wheat, making pythium and rhizoctonia root rots (but not fusarium root and crown rot) largely irrelevant. A 1-ha plot on the WSU Research Station near Lind, WA (in the semi-arid region) and cropped every year to either winter or spring wheat with supplemental sprinkler irrigation (simulating a higher precipitation area) showed maximum take-all by the 7th year and maximum take-all decline by the 15th year (Fig. 1). Thereafter, and through year 19, rhizoctonia root rot caused by *R. solani* AG 8 became increasingly more important, indicating that this otherwise highly destructive root disease was somehow suppressed during the years when take-all was the dominant root disease.

While experimental evidence is limited, the use of certain break crops to control take-all can also leave the soil conducive to the return of take-all with the resumption of wheat monoculture. In the 1-ha long-term wheat-monoculture plot, replicated subplots planted either to alfalfa (lucerne), soybeans, or oats for three consecutive years caused a complete loss of suppressiveness (equivalent to soil fumigation with methyl bromide), compared to replicated subplots of wheat in the same complete block design that maintained suppressiveness (Cook 1981). Two other treatments – replicated plots cropped, respectively, to a mixture of grasses and continuous potatoes – during the 3-years of the experiment were intermediate in maintaining suppressiveness of the soil to take-all when the entire site was again planted uniformly to wheat. Rothrock and Cunfer (1986) reported that take-all decline failed to develop in Georgia in fields double-cropped to wheat and soybeans, suggesting

Fig. 1 Views of the same area within a 1-ha experimental plot cropped to continuous monoculture wheat starting in 1967/68 crop year. *Left*, 1974, the 7th year of monoculture facing north, showing the response to chloropicrin fumigation. *Right*, 1982, the 15th year of monoculture wheat, facing south but otherwise the same area within the 1-ha plot, with the man standing on the border separating a subplot fumigated from an adjacent subplot not fumigated. Yield of wheat in the non-fumigated plots was roughly 50% of the yield in fumigated plots in the 7th year of monoculture and 95% of the yield in fumigated plots in the 15th year of monoculture

that a single crop of soybeans alternated each 12-month period with wheat somehow prevents or limits ability of wheat to enrich and/or maintain a population of PGPR genotypes needed to suppress take-all. The results of Landa et al. (2002) showing that DAPG-producers exceeded log 5.0 CFU/g root of peas grown in monoculture, and that the D and P genotypes were dominant, would suggest that, unlike soybeans, alfalfa and oats in the 1-ha Lind plot, the occasional break to dry peas in the otherwise cereal-intensive cropping systems used in the higher precipitation region of the Inland Northwest may not interfere with the maintenance of take-all suppression.

A large portion of fields in the PNW intermountain region receiving 400 mm or more precipitation annually has produced at least 12 cereal crops in the past 20 years, about half have produced 15–18 and a few fields have produced 20 crops in 20 years. When not planted to a cereal, fields in the higher precipitation areas are typically planted to peas or lentils, and those in the lower precipitation areas are in a 12–14 month fallow alternated with winter wheat. A survey (Ramsey 2001; Cook 2003) of approximately 100 wheat fields in Idaho, Oregon and Washington in each of 1998, 1999, and 2000 using the 0–100 rating system of Shoeny and Lucas (1999) showed that take-all was no more severe in fields cropped every year to wheat (13.3 average rating for 108 fields) or wheat/barley (15.1 average rating for 16 fields) than wheat alternated with a broadleaf (mainly pea or lentil) crop (11.7 average rating for 41 fields). Interestingly, there was no difference in take-all severity between fields direct-seeded (3-year average rating of 13.2) and fields managed with conventional tillage (3-year average rating of 13.5). Raaijmakers and Weller (2001) showed that the population of DAPG-producing pseudomonads in the rhizosphere of wheat in a 1-ha experimental plot near Pullman cropped to wheat or barley for 14 of the previous 16 years was at or above the threshold log 5 CFU/g required for suppression of take-all.

More than conventional systems, the equipment for direct seeding is designed to place nitrogen, phosphorus, and sulphur 3–5 cm beneath or beneath but no farther than 3–5 cm to the side of the seed at the time of planting, thereby placing these nutrients within easy access of the seedling roots. Some growers use a two-pass system, where the bulk of nitrogen is applied through shanks about 10 cm deep prior to sowing and the phosphorus, sulphur and additional nitrogen as a 'starter' mixture is applied with or below the seed at sowing. The practice of applying all nitrogen before or at sowing is in contrast to fertilization practices west of the Cascade Mountains, e.g., in the Willamette Valley of Oregon, and much of Europe, where most or all nitrogen is applied after sowing to avoid leaching. Precision placement of the relatively immobile phosphorous with or below the seed is especially important for crops under pressure from root diseases, to facilitate its access to roots reduced in capacity to explore the soil for this nutrient. Proper placement of phosphorus is also thought to augment the benefits of PGPR in limiting damage from take-all, well known to be more severe on crops starved for phosphorus. The disks or points used to place fertilizer and seed within the same row, by also clearing the residue and creating a narrow zone of tillage within the seed row, is thought to also limit the severity of take-all and pythium and rhizoctonia root rots because of more warming and drying within this zone of disturbed soil. Rhizoctonia root rot caused by *R. solani* AG 8 is particularly sensitive to soil disturbance such as provided by the tools for seed and fertilizer placement (Roget et al. 1996). Paired-row spacing, i.e., positioning two rows 15–20 cm apart with 35–40 cm between the pairs, rather than a uniform 20–30 cm apart, can also help offset the yield-depressing effects of root diseases, by keeping the canopy open longer and adding further to the opportunity for warming and drying of the top few centimeters of soil where the pathogens are most active (Cook et al. 2000).

Control of *Pythium* species starts and ends with the use of fresh high-quality seed (Hering et al. 1987) treated with a chemical protectant (Cook et al. 2002; Smiley et al. 1996). This is especially important for stand establishment in cool wet soils, typical for cereals seeded directly into cereal stubble; emergence of a winter cereal in this system must await the fall rains and emergence of a spring cereal is slow because soil covered with residue of the previous cereal remains cold and wet (Smiley et al. 1996). In addition, fresh wheat straw fragments are highly stimulatory to *Pythium* activity in soil (Cook et al. 1990), resulting in chlorotic and stunted seedings, symptoms that formerly but incorrectly were attributed to phytotoxic substances released from the residue (Cook and Haglund 1991). Metalaxyl (Apron®) is the dominant product used for control of *Pythium* on germinating seeds, used in combination with either difenoconozole

(Dividend®) or tebuconazole (Raxel®) for control of smuts (Smiley et al. 1996). No work has been done in the Pacific Northwest with the seed treatment product fluquinoconazole, shown in Europe to have some benefit against take-all (Bateman et al. 2004), but results from seed-treatment trails conducted with silthiofam, registered in Europe as Latitude® (Beale et al. 1998) and shown to provide some level of take-all control used as seed treatment (Bailey et al. 2005), has shown no benefit in the Pacific Northwest beyond the combination of agricultural practices described above (Cook, unpublished). In contrast, treatment of seeds with metalaxyl combined with either difenoconozole or tebuconazole produced average yield increases of 4–5% in continuous direct-seeded winter and spring wheat using the best agricultural practices described above (Cook et al. 2002).

The ability within the region to grow winter and spring varieties of cereals equally well offers mainly the means to manage weeds but also provides some opportunity to manage rhizoctonia root rot. In spite of the putative reputation of members of this form-genus as soil saprophytes, the rhizoctonias responsible for root rot of cereals depend for their survival primarily on roots of live hosts. Even a relatively short period of no-plants – as short as a few weeks of chemical fallow that includes use of an herbicide to eliminate the volunteer (self-sown) cereals – can result in significantly less pressure from rhizoctonia root rot in the next cereal crop (Roget et al. 1987; Smiley et al. 1992). This discovery has proven particularly important for the management of rhizoctonia root rot of spring-planted cereals where the volunteer cereals and grass weeds are treated with a burn-down herbicide late in the previous fall. Where the volunteer develops too late for fall treatment, spraying as early as possible in the spring and waiting at least 2 weeks and preferably 1 month before planting can also greatly limit the severity of this disease.

Management of fusarium root and crown rot, especially crown rot caused by *F. pseudograminearum*, remains a major challenge in cereal-intensive direct seed systems. Research done in the 1970s on root and crown rot caused by *F. culmorum* showed that the predisposing plant water stress was more likely to develop in crops supplied with excessive nitrogen fertilizer and led to the practice of more carefully matching nitrogen applications to yield goals as set by the amount of water available to finish the crop (Cook 1980). Reducing the amount of nitrogen provided for the crop remains the best if not only practice available to manage this disease. However, this practice has limits for wheat varieties managed for high grain protein, because of the additional nitrogen needed to achieve grain-protein goals.

Conclusions

It is significant that yields within 80–85% of the potential demonstrated in fumigated plots (Cook et al. 2002) can be achieved in cereal-intensive direct-seed cropping systems in spite of the more or less uniform susceptibility of the local cereal cultivars to the four root diseases widely present in these fields. The evidence is clear that DAPG-producing PGPR enriched by intensive cereals and suppressive to take-all is central to the success of these cropping systems. Different cropping systems offer the flexibility to integrate different sequences of spring and winter wheat, barley, and triticale with specific planting and other agricultural practices adapted to local conditions. Fundamental to the success of these systems is the recognition that take-all is only one of at least four different root diseases favoured by direct seeding with intensive cereals, and that dependence exclusively on take-all decline is rarely sufficient under the conditions discussed. Indeed, among take-all and pythium and rhizoctonia root rots, each likely to begin on the seminal roots, and each favoured by the cool moist soil conditions that prevail in direct-seed systems, there is evidence that control of only one can increase the potential for damage caused by one or both of the other two. While this experience shows what can be accomplished through management of resident PGPR as one component of a package approach to design of the cropping system, it does not preclude the potential to augment benefits of resident PGPR through introduction of specific genotypes that might then also be managed for greater performance by use of the cropping system. Even with these gains or potential for gains, the package will not be complete without some progress in breeding for varietal resistance. That may ultimately be the best if not only sustainable solution for control of fusarium root and crown rot.

References

Ahl, P. C., Voisard, C., & Defago, G. (1986). Iron bound-siderophores, cyanic acid, and antibiotics involved in suppression of *Thielaviopsis basicola* by a *Pseudomonas fluorescens* strain. *Journal of Phytopathology, 116*, 121–134.

Bailey, D. J., Paveley, N., Pillinger, C., Foulkes, J., Spink, J., & Gilligan, C. A. (2005). Epidemiology and chemical control of take-all on seminal and adventitious roots of wheat. *Phytopathology, 95*, 62–98.

Bakker, P. A. H. M. (1989). Siderophore-mediated plant growth promotion and colonization of roots by strains of *Pseudomonas* spp. Dissertation, Willie CommelinScholten Phytopathological Laboratory, Department of Plant Pathology, State University Utrecht, Javalaan 20, 3742 Baarn, The Netherlands.

Bangera, M. G., & Thomashow, L. S. (1999). Identification and characterization of a gene cluster for synthesis of the polyketide antibiotic 2,4-diacetylphloroglucinol from *Pseudomonas fluorescens* Q2-87. *Journal of Bacteriology, 181*, 3155–3163.

Bateman, G. L., Gutteridge, R. J., & Jenkyn, J. F. (2004). Take-all and grain yields in sequences of winter wheat crops testing fluquinconazole seed treatment applied in different combinations of years. *Annals of Applied Biology, 145*, 317–330.

Beale R.E., Phillion, D.P., Headrick, J.M., O'Reilly, P., & Cox, J. (1998). MON6550: A unique fungicide for control of take-all in wheat. Paper presented at the 1998 Brighton Conference – Pest and Diseases, British Crop Protection Council Symposium Proceedings, Farnham, Surrey, UK.

Blouin Bankhead, S., Landa, B. B., Lutton, E., Weller, D. M., & McSpadden Gardner, B. B. (2004). Minimal changes in rhizobacterial population structure following root colonization by wild type and transgenic biocontrol strains. *FEMS Microbiology Ecology, 49*, 307–318.

Cook, R. J. (1980). Fusarium foot rot of wheat and its control in the Pacific Northwest. *Plant Disease, 64*, 1061–1066.

Cook, R. J. (1981). The influence of rotation crops on take-all decline phenomenon. *Phytopathology, 71*, 189–192.

Cook, R. J. (1993). Making greater use of introduced micro-organisms for biological control of plant pathogens. *Annual Review of Phytopathology, 31*, 53–80.

Cook, R. J. (2003). Take-all of wheat. *Physiological and Molecular Plant Pathology, 62*, 87–98.

Cook, R. J. (2006). Toward cropping systems that enhance productivity and sustainability. *Proceedings National Academy of Sciences, 103*, 18389–18394.

Cook, R. J. (2007). Take-all decline: A model system in biological control and clue to the success of intensive cropping. In C. Vincent, M. Goettel & G. Lazarovits (Eds.), Biological control a Global Perspective: case studies from around the world. CABI Publishing: UK (in press).

Cook, R. J., & Baker, K. F. (1983). *The Nature and Practice of Biological Control of Plant Pathogens*. St. Paul, MN: American Phytopathological Society.

Cook, R. J., Chamswarng, C., & Tang, W.-H. (1990). Influence of wheat chaff and tillage on *Pythium* populations and *Pythium* damage to wheat. *Soil Biology Biochemistry, 22*, 939–947.

Cook, R. J., & Haglund, W. A. (1991). Wheat yield depression associated with conservation tillage caused by root pathogens in the soil not phytotoxins from the straw. *Soil Biology Biochemistry, 23*, 1125–1132.

Cook, R. J., Ownley, B. H., Zhang, H., & Vakoch, D. (2000). Influence of paired-row spacing and fertilizer placement on yield and root diseases of direct-seeded wheat. *Crop Science, 40*, 1079–1087.

Cook, R. J., & Rovira, A. D. (1976). The role of bacteria in the biological control of Gaeumannomyces graminis by suppressive soils. *Soil Biology Biochemistry, 8*, 269–273.

Cook. R. J., Thomashow, L. S., Weller, D. M., Fujimoto, D. K., Mazzola, M., Bangera, G., et al. (1995). Molecular mechanisms of defense by rhizobacteria against root disease. *Proceedings National Academy Sciences, 92*, 4197–4201.

Cook, R. J., & Veseth, R. J. (1991). *Wheat Health Management*. St. Paul, MN: APS.

Cook, R. J., Weller, D. M., El-Banna, A. Y., Vakoch, D., & Zhang, H. (2002). Yield responses of direct-seeded wheat to fungicide and rhizobacteria seed-treatments. *Plant Disease, 86*, 780–784.

Fenton, A. M., Stephens, P. M., Crowley, J., O'Callaghan, M., & O'Gara, F. (1992). Exploitation of gene(s) involved in 2,4-diacetylphloroglucinol biosynthesis to confer a new biocontrol capability to a *Pseudomonas* strain. *Applied Environmental Microbiology, 58*, 3873–3878.

Gerlach, M. (1968). Introduction of *Ophiobolus graminis* into new polders and its decline. *Netherlands Journal of Plant Pathology, 74* (Supplement 2), 1–97.

Harman, G. E., Howell, C., Viterbo, A, Chet, I., & Matteo, L. (2004). *Trichoderma* species–Opportunistic, avirulent plant symbionts. *Nature Reviews Microbiology, 2*, 43–56.

Hass, D., & Defago, G. (2005). Biological control of soil-borne pathongens by fluorescent pseudomonads. *Nature Reviews Microbiology, 3*, 307–319.

Hering, T. F., Cook, R. J., & Tang, W.-H. (1987). Infection of wheat embryos by *Pythium* species during seed germination and the influence of seed age and soil matric potential. *Phytopathology, 77*, 1104–1108.

Keel, C., Schnider, U., Maurhofer, M., Voisard, C., & Burger, D. (1992). Suppression of root diseases by *Pseudomonas fluorescens* CHAO: Importance of the bacterial secondary metabolite 2,4-diacetylphloroglucinol. *Molecular Plant-Microbe Interactions, 5*, 4–13.

Keel, C., Weller, D. M., Natsch., A., Défago, G., Cook R. J., & Thomashow L. S. (1996). Conservation of the 2,4-diacetyl-phloroglucinol biosynthesis locus among fluorescent *Pseudomonas* strains from diverse geographic locations. *Applied and Environmental Microbiology, 62*, 552–563.

Landa, B. B., Mavrodi, O. V., Raaijmakers, J. M., McSpadden Gardener, B. B., Thomashow, L. S., & Weller, D. M. (2002). Differential ability of genotypes of 2,4-diacetyl-phloroglucinol-producing *Pseudomonas fluorescens* strains to colonize the roots of pea plants. *Applied and Environmental Microbiology, 68*, 3226–3237.

Landa, B. B., Mavrodi, O., Schroeder, K., Allende-Molar, R., & Weller, D. M. (2005). Enrichment and genotypic diversity of *phl-D* containing fluorescent *Pseudomonas* spp. in two

soils after a century of wheat and flax monoculture. *FEMS Microbiology Ecology, 55*, 351–356.
Mathre, D. M., Cook, R. J., & Callan, N. W. (1999). From discovery to use: Traversing the world of commercializing biocontrol agents for plant disease control. *Plant Disease, 83*, 972–983.
Mavrodi, D. V., Ksenzenko, V. N., Bonsall, R. F., Cook, R. J., Boronin, A. M., & Thomashow, L. S. (1998). A seven-gene locus for synthesis of phenazine-1-carboxylic acid by *Pseudomonas fluorescens* 2–79. *Journal Bacteriology, 180*, 2541–2548.
Mavrodi, O. V., McSpadden Gardener, B. B., Mavrodi, D. V., Bonsall, R. F., Weller, D. M., & Thomashow, L. S. (2001). Genetic diversity of *phlD* from 2,4-diacetylphloroglucinol-producing fluorescent *Pseudomonas* species. *Phytopathology, 91*, 35–43.
Mazzola, M. (1998). Elucidation of the microbial complex having a causal role in the development of apple replant disease in Washington. *Phytopathology, 88*, 930–938.
Mazzola, M. (1999). Transformation of soil microbial community structure and *Rhizoctonia*-suppressive potential in response to apple roots. *Phytopathology, 89*, 920–927.
Mazzola, M., & Gu, Y.-H. (2000). Phytomanagement of microbial community structure to enhance growth of apples in replant soils. *Acta Horticulture, 532*, 73–78.
McSpadden Gardener, B. B., Mavrodi, D. V., Thomashow, L. S., & Weller, D. M. (2001). A rapid PCR-based assay characterizing rhizosphere populations of 2,4-DAPG-producing bacteria. *Phytopathology, 91*, 44–54.
McSpadden Gardener, B. B., Schroeder, K. L., Kalloger, S. E., Raaijmakers, J. M., Thomashow, L. S., & Weller, D. M. (2000). Genotypic and phenotypic diversity of *phl-D*-containing *Pseudomonas* isolated from the rhizosphere of wheat. *Applied Environmental Microbiology, 66*, 1939–1946.
McSpadden Gardner, B. B., Gutierrez, L. J., Raghavendra, J., Edema, R., & Lutton, E. (2005). Distribution and biocontrol potential of *phlD*$^+$ pseudomonads in corn and soybean fields. *Phytopathology, 95*, 15–724.
Paulitz, T., Smiley, R. W., & Cook, R. J. (2002). New insights into the make-up and management of soilborne cereal pathogens under direct seeding in the Pacific Northwest. *Canadian Journal of Plant Pathology, 24*, 416–428.
Picard, C., Di Cello, F., Ventura, M., Fani, R., & Guckert, A. (2000). Frequency and biodiversity of 2,4-diacetylphloroglucinol-producing bacteria isolated from the maize rhizosphere at different stages of plant growth. *Applied Environmental Microbiology, 66*, 948–955.
Raaijmakers, J. M., Bonsall, R. F., & Weller, D. M. (1999). Effect of population density of *Pseudomonas fluorescens* on production of 2,4-diacetylphloroglucinol in the rhizosphere of wheat. *Phytopathology, 89*, 470–475.
Raaijmakers, J. M., & Weller, D. M. (2001). Exploiting genotypic diversity of 2,4-diacetylphloroglucinol-producing *Pseudomonas* spp: Characterization of superior root-colonizing *P. fluorescens* strain Q8r1-96. *Applied Environmental Microbiology, 67*, 2545–2554.
Ramsey, N. E. (2001). Occurrence of take-all on wheat in Pacific Northwest cropping systems. Pullman, WA: Dissertation, Washington State University.
Rassmusen, P. E., & Rhode, C. R. (1989). Longer-term tillage and nitrogen fertilization effects on organic nitrogen and carbon in a semiarid soil. *Soil Science Society of America Journal, 53*, 119–122.
Roget, D. K., Neate, S. M., & Rovira, A. D. (1996). Effect of sowing point design and tillage practice on the incidence of rhizoctonia root rot, take-all and cereal cyst nematode in wheat and barley. *Australian Journal Experimental Agriculture, 36*, 683–693.
Roget, D. K., Venn, N. R., & Rovira, A. D. (1987). Reduction of rhizoctonia root rot of direct-drilled wheat by short-term chemical fallow. *Australian Journal of Experimental Agriculture, 27*, 425–430.
Rothrock, C. S., & Cunfer, B. M. (1986). Absence of take-all decline in double-cropped fields. *Soil Biology and Biochemistry, 18*,113–114.
Sands, D. C., & Rovira, A. D. (1970). Isolation of fluorescent psuedomonads with a selective medium. *Applied Microbiology, 20*, 513–514.
Schippers, B., Bakker, A. W., & Bakker, P. A. H. M. (1987). Interactions of deleterious and beneficial rhizosphere microorganisms and the effect of cropping practices. *Annual Review of Phytopathology, 25*, 258–339.
Schoeny, A., & Lucas, P. (1999). Modeling of take-all epidemics to evaluate the efficacy of a new seed-treatment fungicide on wheat. *Phytopathology, 89*, 954–961.
Shipton, P. J., Cook, R. J., & Sitton, J. W. (1973). Occurrence and transfer of a biological factor in soil that suppresses take-all in wheat in eastern Washington. *Phytopathology, 63*, 511–517.
Smiley, R. W., Merrifield, K., Patterson, L. M., Gourlie, J. A., & Easley, S. A. (2004). Nematodes in dryland field crops in the semiarid Pacific Northwest USA. *Journal of Nematology, 36*, 54–68.
Smiley, R. W., Ogg, Jr., A. G., & Cook, R. J. (1992). Influence of glyphosate on severity of Rhizoctonia root rot and growth and yield of barley. *Plant Disease, 76*, 937–942.
Smiley, R. W., Patterson, L. M., & Sheldon, C. W. (1996). Fungicide seed treatments influence emergence of winter wheat in cold soil. *Journal of Production Agriculture, 9*, 559–563.
Souza, J. T., Weller, D. M., & Raaijmakers, J. M. (2003). Frequency, diversity, and activity of 2,4-diacetylphloroglucinol-producing fluorescent *Pseudomonas* spp. in Dutch take-all decline soils. *Phytopathology, 93*, 54–63.
Stutz, E. W., Defago, G., & Kern, H. (1986). Naturally occurring fluorescent pseudomonads involved in suppression of black root rot of tobacco. *Phytopathology, 76*, 181–185.
Weller, D. M. (1988). Biological control of soilborne plant pathogens in the rhizosphere by bacteria. *Annual Review of Phytopathology, 26*, 379–407.
Weller, D. M., Landa, B. B., Mavrodi, O. V., Schroeder, K. L., De La Fuente, L., Blouin Bankhead, S., et al. (2007). Role of 2,4–diacetylphloroglucinol-producing fluorescent *Pseudomonas* spp. in the defense of plant roots. *Plant Biology, 9*, 4–20.
Weller, D. M., Raaijmakers, J. M., McSpadden Gardener, B. B., & Thomashow, L. S. (2002). Microbial populations responsible for specific soil suppressiveness to plant pathogens. *Annual Review of Phytopathology, 40*, 309–348.

Eur J Plant Pathol (2007) 119:265–278
DOI 10.1007/s10658-007-9179-8

REVIEW PAPER

Genomic analysis of antifungal metabolite production by *Pseudomonas fluorescens* Pf-5

Joyce E. Loper · Harald Gross

Received: 18 December 2006 / Accepted: 24 May 2007 / Published online: 5 July 2007

Abstract The complete genomic sequences of several *Pseudomonas* spp. that inhabit the rhizosphere are now available, providing a new opportunity to advance knowledge of plant growth-promoting rhizobacteria (PGPR) through genomics. Among these is the biological control bacterium *Pseudomonas fluorescens* Pf-5. Nearly 6% of the 7.07 Mb genome of Pf-5 is devoted to the biosynthesis of secondary metabolites, including antibiotics toxic to soilborne fungi and Oomycetes that infect plant roots, and two siderophores involved in iron acquisition. Three orphan gene clusters, for which the encoded natural product was unknown, also were identified in the genome of Pf-5. The product synthesized from one of the orphan gene clusters was identified recently using a new 'genomisotopic approach', which employs a combination of genomic sequence analysis and isotope guided fractionation. Application of the genomisotopic approach to one orphan gene cluster in Pf-5 resulted in the discovery of orfamide A, founder of a new group of bioactive cyclic lipopeptides with a putative role in biological control of plant disease.

Keywords Cyanide · Cyclic lipopeptides · 2,4-Diacetylphloroglucinol · Orphan gene clusters · Pyrrolnitrin · Pyoluteorin · Mcf toxin

Abbreviations

CLP	cyclic lipopeptide
DAPG	2,4-diacetylphloroglucinol
GI approach	genomisotopic approach
HCN	hydrogen cyanide
NRPS	non-ribosomal peptide synthetase
PGPR	Plant growth-promoting rhizobacteria
PKS	polyketide synthase
Mcf	'makes caterpillars floppy'

J. E. Loper (✉)
Horticultural Crops Research Laboratory, Agricultural Research Service, US Department of Agriculture, 3420 N.W. Orchard Ave., Corvallis, OR, USA
e-mail: loperj@science.oregonstate.edu

H. Gross
Institute for Pharmaceutical Biology, University of Bonn, Nussallee 6, 53115 Bonn, Germany

Introduction

Plant growth-promoting rhizobacteria (PGPR) are a diverse group of organisms that share two characteristics: the capacity to colonize the rhizosphere and to have a positive influence on the growth of plants whose rhizospheres they inhabit. Knowledge of these rhizobacteria and their interactions with plants and other components of microbial communities associated with roots has advanced tremendously during the past two decades, and molecular approaches have been important tools employed to build this knowledge. With the recent availability of genomic sequence data for rhizosphere bacteria, the field of genomics can now contribute to advancing knowledge of PGPR and their effects on plant health and productivity. Complete genomic sequences for five species of *Pseudomonas*, including several strains that colonize the rhizosphere, are available to date (December 2006). This review will focus on the rhizosphere bacterium *Pseudomonas fluorescens* Pf-5, whose complete genome was sequenced recently (Paulsen et al. 2005). Here, we provide a brief summary of the biology of Pf-5 and its genome, before focusing on gene clusters for antifungal metabolites and the characterization of Orfamide A, a newly discovered cyclic lipopeptide (CLP) whose structure was predicted from the genomic sequence data and solved using a new genomisotopic approach. We also refer the reader to recent articles providing further analysis of the genomic sequence of Pf-5 (Loper et al. 2007; Mavrodi et al. 2007).

Pseudomonas fluorescens strain Pf-5

Strain Pf-5, which was isolated from the soil in Texas, USA, was first described for its capacity to suppress soilborne diseases of cotton caused by *Rhizoctonia solani* (Howell and Stipanovic 1979) and *Pythium ultimum* (Howell and Stipanovic 1980). Pf-5 has since been shown to suppress these pathogens on other plant hosts including cucumber (Kraus and Loper 1992) and pea (M. D. Henkels and J. Loper, unpublished). Pf-5 also suppresses a number of other soilborne or residue-borne fungal pathogens. When inoculated onto wheat straw residue, Pf-5 suppresses ascocarp formation by the tan spot pathogen of wheat, *Pyrenophora tritici-repentis* (Pfender et al. 1993). The biocontrol agent also suppresses two diseases of turf grass: dollar spot caused by *Sclerotinia homoeocarpa* and a leaf spot caused by *Dreschlera poae* (Rodriguez and Pfender 1997). These are widespread, destructive diseases affecting golf courses, home lawns, and amenity turf areas. Pf-5 also suppresses Fusarium root and crown rot of tomato, caused by *Fusarium oxysporum* f. sp. *radicis-lycopersici* (Sharifi-Tehrani et al. 1998) and seed piece decay of potato caused by the bacterial pathogen *Erwinia carotovora* (Xu and Gross 1986).

Secondary metabolite production by Pf-5

Pf-5 produces the antibiotics pyrrolnitrin (Howell and Stipanovic 1979), pyoluteorin (Howell and Stipanovic 1980; Kraus and Loper 1995), and 2,4-diacetylphloroglucinol (Nowak-Thompson et al. 1994); it also produces hydrogen cyanide (Kraus and Loper 1992) and two siderophores: a pyoverdine of unconfirmed structure and pyochelin (or a related compound). The spectrum of antibiotics produced by Pf-5 is remarkably similar to that produced by the well-characterized biological control strain *P. fluorescens* CHA0, which was isolated from roots of tobacco grown in a soil near Payerne, Switzerland (Stutz et al. 1986; Haas and Keel 2003; Haas and Défago 2005). Many other biological control strains of rhizobacteria produce a subset of metabolites produced by Pf-5 and CHA0 (Raaijmakers et al. 2002). Pf-5 does not produce the phenazine antibiotics that are produced by certain biological control strains of *Pseudomonas* spp. (Mavrodi et al. 2006).

The genomic sequence of Pf-5

In this section, we provide a brief overview of the genome of Pf-5, focusing on aspects of the genomic sequence of particular significance to the ecology of this rhizosphere bacterium and its interactions with the plant and other plant-associated microorganisms. Data presented in the following paragraphs are summarized from Paulsen et al. (2005).

Pseudomonas fluorescens Pf-5 inhabits root and seed surfaces, and its genome has a complement of genes specifying broad metabolic capacity, including the utilization of a variety of organic acids, sugars, and amino acids found in seed or root exudates. In this respect, it is very similar to *P. putida* KT2440, which also inhabits the rhizosphere (dos Santos et al. 2004). Pf-5 has genes for the metabolism of plant-derived carbohydrates such as maltose, sucrose, trehalose and xylose, and for more complex plant-derived molecules such as the aromatic compounds vanillate, benzoate and hydroxybenzoate, as well as long chain fatty acids and hydrocarbons, of which many plant oils are comprised. Present in its genome are several extracellular hydrolytic enzymes, including chitinases, proteases and lipases, which are involved in the degradation of polymers commonly found in soil. For iron acquisition, the genome specifies the biosynthesis of two siderophores, as well as 45 predicted TonB-dependent outer membrane proteins, many of which are likely to serve as receptors for a diverse collection of ferric siderophores. *Pseudomonas* spp. are known to utilize siderophores produced by other microorganisms as sources of iron. In natural habitats on root surfaces, these bacteria can acquire iron by uptake of exogenous siderophores, obviating the need to rely on siderophore production alone (Loper and Henkels 1999). Determining the roles of these outer membrane receptors in the ecology of Pf-5 in the soil will be an enlightening subject for future inquiry.

Bacteria inhabiting the rhizosphere are likely to be exposed to a variety of toxic metabolites produced by the plant or other microorganisms. The genome of Pf-5 contains an expanded set of membrane efflux systems, which typically confer protection against a range of toxic metabolites. Among those are genes with predicted roles in the efflux of secondary metabolites produced by Pf-5 (Abbas et al. 2004; Brodhagen et al. 2005; Huang et al. 2006), resistance to fusaric acid and copper (Mellano and Cooksey 1998), and the phytotoxin tabtoxin. Consistent with the proposed importance of oxidative stress tolerance as a determinant of rhizosphere fitness, the genome encodes multiple peroxidases, catalases, and superoxide dismutases.

As expected of an organism living in a complex and rapidly changing environment, Pf-5 has an extensive array of regulatory genes. These include 68 predicted histidine kinases and 113 predicted response regulators (Kiil et al. 2005b). Pf-5 has 27 genes encoding sigma factors in the extracytoplasmic factor (ECF) class, the most among all of the *Proteobacteria* whose genomes have been sequenced (Kiil et al. 2005a). These sigma factors typically coordinate transcriptional responses to extracellular signals and have diverse functions in iron acquisition, stress response, metal resistance, cell development, virulence, and the production of extracellular products (Helmann 2002). In Pf-5, 18 of the 27 ECF sigma factor genes are adjacent to genes encoding predicted TonB-dependent outer-membrane proteins, indicating a possible role in iron acquisition (Mavrodi et al. 2007). The numerous genes with putative roles in transcriptional regulation indicate that exceedingly complex regulatory networks exist in this environmental bacterium.

The genome of Pf-5 is notable for the absence of virulence factors found in pathogenic *Pseudomonas* spp. Lacking are genes for the synthesis of the phytotoxins tabtoxin, syringomycin, syringotoxin, syringopeptin, or coronatine. Also absent are genes for exoenzymes associated with degradation of plant cell walls or cell wall components, such as amylases or cellulases. There is no evidence for a type III secretion system although genes for such systems have been reported in many other strains of *P. fluorescens* (Mazurier et al. 2004; Preston et al. 2001; Rezzonico et al. 2005).

Pf-5 has several genes that specify exported factors with possible roles in biological control. The genome contains a homolog of *chiC* (Folders et al. 2001), which encodes chitinase, an enzyme that degrades chitin, an important fungal cell wall component. Chitinase production by other microorganisms has been implicated in biological control (Harman et al. 2004), and Pf-5 produces the enzyme (Donald Kobayashi, personal communication) but its role in disease suppression by Pf-5 has not been established. As typical for the genus, Pf-5 produces an extracellular alkaline protease(s) and has two homologs of the exoprotease gene *aprA*, which contributes to root knot nematode suppression by *P. fluorescens* CHA0 (Siddiqui et al. 2005). The genome also has two homologs of *llpA*, which encodes a bacteriocin related to LlpA (Parret et al. 2005). Intriguingly, a homolog of *mcf* (for makes caterpillars floppy) is also present in the Pf-5 genome. Mcf is an insect toxin produced by the bacterium *Photorhabdus luminescens*, an inhabitant of the gut of entomopathogenic nematodes (Daborn et al. 2002). If injected into the hemocoel, Pf-5 kills caterpillars of tobacco hornworms (*Manduca sexta*) whereas an *mcf* mutant of Pf-5 is less virulent (M. Pechy-Tarr, D. Bruck, M. Maurhofer, M. Henkels, K. Donahue, J. Loper, and C. Keel, unpublished data). The genomic sequence data provides direction for new inquiries highlighting the diverse biological activities of this fascinating soil bacterium.

Antibiotic biosynthesis gene clusters

Antibiotics produced by rhizosphere bacteria play an important role in biological control, and several excellent reviews provide perspectives of the recent literature on this subject (Haas and Keel 2003; Raaijmakers et al. 2002; Mavrodi et al. 2006; Weller et al. 2002). Well-characterized antibiotics that contribute to biological control include pyoluteorin, 2,4-diacetylphloroglucinol, pyrrolnitrin, and hydrogen cyanide, all of which are produced by *P. fluorescens* Pf-5. Nearly 6% of the Pf-5 genome is devoted to secondary metabolism, with loci for nine different secondary metabolites distributed around the genome (Fig. 1). Two of the gene clusters are involved in siderophore biosynthesis and two contain biosynthetic loci for which the corresponding metabolite is unknown. The following discussion will focus on the five gene clusters that contain biosynthetic loci

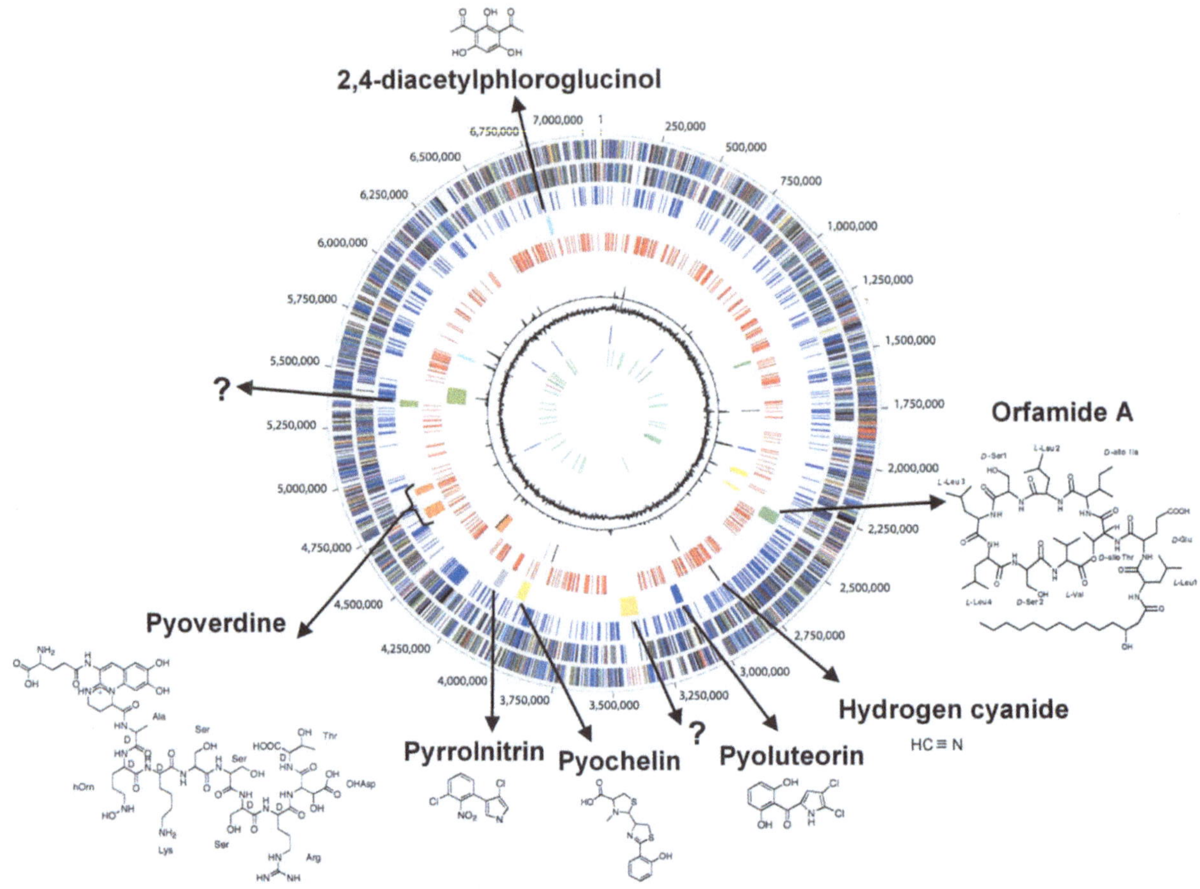

Fig. 1 Circular representation of the genome of *Pseudomonas fluorescens* Pf-5. The outer scale designates coordinates in base pairs (bp), with the origin of replication at 1 bp. The first circle (*outermost circle*) shows predicted coding regions on the plus strand colour-coded by role categories: violet, amino acid biosynthesis; light blue, biosynthesis of cofactors, prosthetic groups and carriers; light green, cell envelope; red, cellular processes; brown, central intermediary metabolism; yellow, DNA metabolism; light grey, energy metabolism; magenta, fatty acid and phospholipid metabolism; pink, protein synthesis and fate; orange, purines, pyrimidines, nucleosides and nucleotides; olive, regulatory functions and signal transduction; dark green, transcription; teal, transport and binding proteins; grey, unknown function; salmon, other categories; blue, hypothetical proteins. The second circle shows predicted coding regions on the minus strand colour-coded by role categories. The third circle shows the set of 656 *P. fluorescens* Pf-5 genes not found in the genomes of *P. aeruginosa* PAO1, *P. syringae* pv. tomato DC3000, and *P. putida* KT2440. The fourth circle shows nine secondary metabolite gene clusters, with the structures and names of the corresponding metabolite indicated with lines. Two orphan gene clusters, whose metabolic products are unknown, are designated with a question mark. The fifth circle shows 1,052 copies of a 34 bp REP repeat element. The sixth circle shows a mobile island in olive and seven phage regions. The seventh circle shows trinucleotide composition. The eighth circle shows percentage G + C in relation to the mean G + C in a 2,000-bp window. The ninth circle shows rRNA operons in blue. The tenth circle (*innermost circle*) shows tRNA genes in green. Adapted from Paulsen et al. (2005)

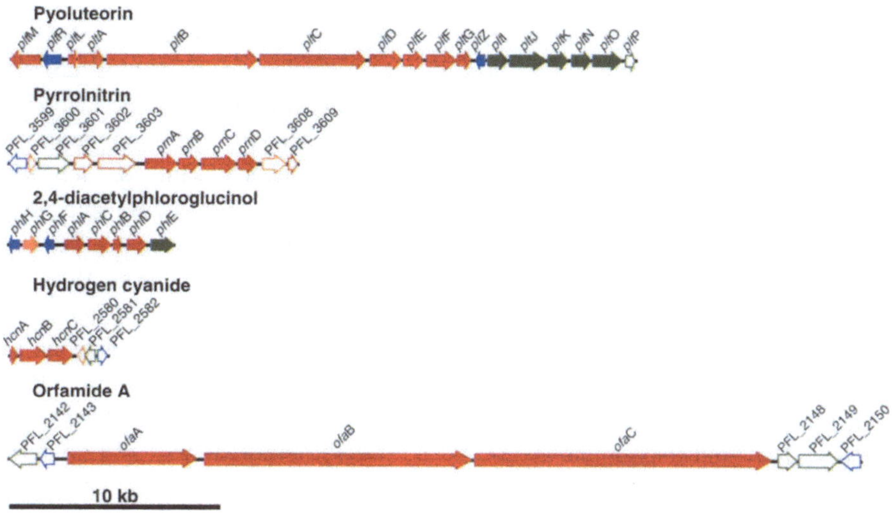

Fig. 2 Antibiotic biosynthetic gene clusters in the genome of *P. fluorescens* Pf-5. Solid arrows denote genes for which there is experimental evidence for involvement in antibiotic production by *Pseudomonas* spp. Red, structural genes for antibiotic biosynthesis; blue, regulation; green, transport; orange, accessory genes. Open arrows denote genes flanking antibiotic biosynthetic gene clusters and are shown to provide context within the genome of Pf-5. Colours of open arrows correspond to the same deduced functions defined for solid arrows. Adapted from Paulsen et al. (2005)

for antibiotics with known structures, one of which has been identified using the Pf-5 genomic sequence data as a starting point.

Pyoluteorin

Pyoluteorin is toxic against the Oomycete *Pythium ultimum* and certain other soilborne pathogens (Howell and Stipanovic 1980; Maurhofer et al. 1995). Pyoluteorin also exhibits some phytotoxicity (Maurhofer et al. 1995; Rao and Reddy 1990). The sequence of the pyoluteorin biosynthetic gene cluster was first described in strain Pf-5 (Nowak-Thompson et al. 1997, 1999). Genes functioning in regulation (*pltZ* and *pltR*) and efflux of pyoluteorin (*pltIJKLMO*) are present in the biosynthetic gene cluster (Fig. 2) (Brodhagen et al. 2005; Huang et al. 2004, 2006; Nowak-Thompson et al. 1999). The organization of the pyoluteorin biosynthesis and efflux gene cluster of *Pseudomonas* sp. M18 (Huang et al. 2006) is identical to that in Pf-5, although the nomenclature used for the efflux genes in M18 is slightly different from that used in Pf-5.

2,4-Diacetylphloroglucinol (DAPG)

DAPG is toxic to a wide range of plant pathogenic fungi and also exhibits antibacterial and anthelminthic properties (Cronin et al. 1997; Keel et al. 1992). At relatively high concentrations, it is phytotoxic (Keel et al. 1992). Many strains of *P. fluorescens* produce DAPG, and phylogenetic analysis of the DAPG-producing strains places them in at least 22 genotypes (Weller et al. 2007), with Pf-5 classified in the A genotype of DAPG producers. Different genotypes of DAPG producers predominate in different soils, but the D genotype of DAPG producers predominates in the rhizosphere of wheat in soils suppressive to take-all decline, the most well characterized system exhibiting natural biological control (Weller et al. 2002, 2007).

As already demonstrated in many other strains of *P. fluorescens*, genes for DAPG biosynthesis (*phlACB* and *phlD*) (Bangera et al. 1999), efflux (*phlE*) (Abbas et al. 2004), and degradation (*phlG*) (Bottiglieri and Keel 2006) are clustered with those for regulation (*phlF* and *phlH*) (Abbas et al. 2002; Delany et al. 2000; Schnider-Keel et al. 2000) in the genome of Pf-5 (Fig. 2). The organization of the gene cluster is identical and the predicted amino acid sequences of the genes are also highly conserved (between 71 and 99% identity) with homologs in other strains.

Pyrrolnitrin

A diverse range of fungi, representing Basidiomycetes, Deuteromycetes, and Ascomycetes, are sensitive to pyrrolnitrin (Ligon et al. 2000). Synthetic analogues of pyrrolnitrin have been developed as agricultural fungicides (Ligon et al. 2000), and pyrrolnitrin has also been used as a topical antibiotic for fungal infections of humans. The antibiotic is an inhibitor of fungal respiratory chains (Tripathi and Gottlieb 1969).

Four biosynthetic genes (*prnABCD*) are highly conserved among strains of *P. fluorescens* that produce pyrrolnitrin (Hammer et al. 1997, 1999; Kirner et al. 1998). Expression of *prnABCD* in heterologous bacteria results in pyrrolnitrin production (Hammer et al. 1997), indicating that these genes are sufficient for biosynthesis of the antibiotic from primary precursors. The predicted amino acid sequences of the four genes in the Pf-5 genome (Fig. 2) are very similar (between 90 and 99% identity) to those of homologs in *P. fluorescens* strains CHA0 (Baehler et al. 2005) and BL915 (Hammer et al. 1999). The amino acid sequences of the four genes are also very similar to homologs in pyrrolnitrin-producing strains of *Burkholderia* spp. such as *B. cepacia* strain AMMD (87–95% identity) (accession # NC_008391). Flanking the four known biosynthetic genes in the Pf-5 genome are other genes that could possibly have a role in regulation, efflux, or production of pyrrolnitrin (Fig. 2). Especially notable is PFL_3609, a member of the flavin reductase family, which could provide the reduced $FADH_2$ required for enzymatic activity of the halogenating enzymes PrnA and PrnC (Dong et al. 2005). Other genes of interest include PFL_3601, a transporter in the major facilitator subfamily (MFS), which commonly function in efflux; and PFL_3599, a transcriptional regulator in the LysR family. Because pyrrolnitrin is produced by heterologous bacteria such as *Escherichia coli* harbouring the *prnABCD* operon alone (Hammer et al. 1997), the flanking genes are not strictly required for pyrrolnitrin biosynthesis. The potential roles of these genes in pyrrolnitrin production by Pf-5 are subjects for future exploration.

Hydrogen cyanide

HCN is an effective inhibitor of cytochrome c oxidase (Knowles 1976) and other metalloenzymes (Blumer and Haas 2000). Its production by PGPR is implicated in biological control of black root rot of tobacco (Voisard et al. 1989), root rot of tomato, and Pythium damping-off of cucumber (Ramette et al. 2003). Cyanogenesis by *Pseudomonas* spp. is the product of an HCN synthase, encoded by three biosynthetic genes (*hcnABC*) (Laville et al. 1998). These three genes are sufficient for HCN biosynthesis from glycine, the primary precursor (Blumer and Haas 2000).

The *hcnABC* operon in the Pf-5 genome (Fig. 2) is virtually identical in sequence (99–100% identity at the amino acid level) and organization to the corresponding HCN biosynthetic genes in *P. fluorescens* CHA0 (Laville et al. 1998), which conforms to an earlier comparison of *hcnBC* among cyanogenic *Pseudomonas* spp. (Ramette et al. 2003). Immediately downstream of the *hcnABC* operon is PFL_2580, a member of the glutathione S-transferase (GST) family, a large and diverse group of proteins that function in detoxification of a wide range of endogenous and xenobiotic compounds. Homologs of PFL_2580 are found in the genomes of the cyanogenic species *P. aeruginosa* and *Pseudomonas entomophila*, but they are not linked to *hcnABC* in those species. Multiple mechanisms contribute to HCN resistance in *P. aeruginosa*. These include *rhdA*, which encodes a rhodonase that detoxifies HCN (Cipollone et al. 2007); the *cioAB* gene pair, which encode a cyanide insensitive terminal oxidase (Cooper et al. 2003); and the excretion of α-ketoglutarate, which detoxifies cyanide (Blumer and Haas 2000). As in *P. aeruginosa*, homologs of *cioAB* and *rhdA* are not linked to the *hncABC* biosynthetic locus in the Pf-5 genome. Two other ORFs neighbouring *hcnABC* are PFL_2581, a member of the MFS superfamily of membrane transporters, and PFL_2582, a transcriptional regulator. Further study is needed to ascertain the potential roles of these genes in hydrogen cyanide production by Pf-5.

Orphan gene clusters

An exciting outcome of the Pf-5 genomic sequencing project was the identification of three orphan gene clusters. These three orphan gene clusters contain sequences that are characteristic of polyketide synthases (PKSs) or non-ribosomal peptide synthetases (NRPSs), enzymes that catalyze the formation of secondary metabolites through a non-ribosomal mechanism of biosynthesis (Wenzel and Müller 2005). Although the structures of compounds that are generated non-ribosomally are diverse, their biosynthetic enzymes share a modular architecture of catalytic domains and an assembly line mechanism of product synthesis. Furthermore, the catalytic domains are encoded by highly conserved sequences that can be used to identify novel biosynthetic gene clusters containing PKSs or NRPSs. The three orphan gene clusters were found by searching the Pf-5 genomic sequence data for the conserved sequences of PKSs and NRPSs. To date, our efforts to identify metabolites synthesized from the orphan gene clusters of Pf-5 have focused on one cluster containing a NRPS.

NRPSs are large, multifunctional enzymes that are organized into modules, each of which synthesizes the addition of an amino acid to a peptide, which is then transferred to downstream modules for the next step in the synthesis of an elongating peptide chain (Finking and Marahiel 2004). Each elongation module contains at least three types of domains: (i) a condensation (C) domain that catalyzes peptide bond formation between amino acids in the growing peptide chain; (ii) an adenylation (A) domain, which is responsible for recognition and activation of a specific amino acid; and (iii) a thiolation (T) domain, which binds the substrate as it moves from module to module along the NRPS. Finally, the elongation process is terminated by the release of the final product through hydrolysis or cyclization, the latter process often catalyzed by a thioesterase (TE).

A remarkable characteristic of natural product biosynthesis via NRPSs is that the number and order of the modules of an NRPS protein generally are colinear to the amino acid sequence of the final peptide product, a relationship termed the 'colinearity rule.' Furthermore, the modules contain characteristic motifs, primarily in the A domains, that select for a specific amino acid to be added to the growing peptide chain (Stachelhaus et al. 1999). These signature motifs can be identified using web-based bioinformatic algorithms (Ansari et al. 2004; Challis et al. 2000) and, because of the colinearity rule, they can be used to predict the order, composition and number of amino acids in the final peptide product.

Below, we summarize the process by which one of the orphan metabolites in the Pf-5 genome was purified and characterized, using the genomic sequence data as a starting point.

Sequence analysis of an orphan gene cluster in the genome of Pf-5

One of the three orphan gene clusters in the Pf-5 genome (Fig. 3a) contained three contiguous genes, termed *ofa*A, *ofa*B, and *ofa*C, whose deduced amino acid sequences are similar to NRPSs (Gross et al. 2007; Paulsen et al. 2005). Together, OfaA, OfaB, and OfaC comprise ten modules, each of which contains a condensation (C), adenylation (A), and thiolation (T) domain characteristic of NRPS modules (Fig. 3b). In addition to the ten modules, two thioesterase (TE) domains, required for liberation of the peptide from the enzyme complex, are present in the C-terminus of the predicted amino acid sequence of OfaC. Analysis of the DNA sequence of *ofaA*, *ofaB*, and *ofaC* provided three lines of evidence that the product of the NRPS is a cyclic lipopeptide (CLP), a large class of compounds composed of a fatty acid tail linked to a peptide that is cyclized to form a lactone ring (Nybroe and Sørensen 2004; Raaijmakers et al. 2006): (1) the predicted amino acid sequences of OfaA, OfaB, and OfaC are most related to NRPSs for the biosynthesis of CLPs by other *Pseudomonas* species, (2) the Ofa NRPS lacks a typical initiation module, which is also lacking in the NRPSs for other CLPs (Guenzi et al. 1998; Roongsawang et al. 2003; Scholz-Schroeder et al. 2003); while initiation modules for NRPSs typically lack a C domain, the first module in the NRPS for CLP biosynthesis contains a C domain, which is likely to be involved with bond formation with the lipid starter unit (Roongsawang et al. 2003), (3) two TE domains are present in the C-terminus of the Ofa NRPS, a characteristic of NRPSs for CLPs of *Pseudomonas* spp.

The amino acid sequence of the peptide component of the orphan metabolite was predicted through analysis of signature sequences within the A domains of OfaA, OfaB, and OfaC (Gross et al. 2007; Paulsen et al. 2005). This analysis provided further evidence that the orphan metabolite is a CLP, because the deduced amino acid sequence of peptidyl product is very similar to those of several known CLPs (Gross et al. 2007; Paulsen et al. 2005). Nevertheless, the orphan metabolite from Pf-5 is distinct from previously-described CLPs produced by *Pseudomonas* spp., which are classified into five groups according to the

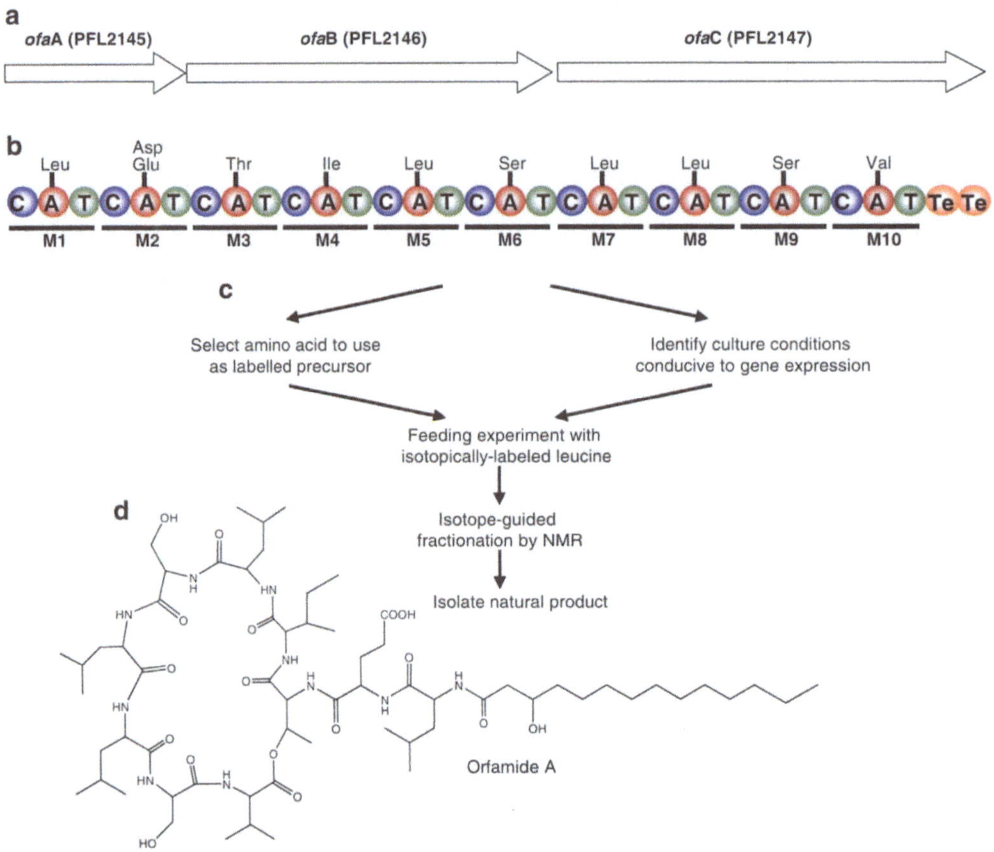

Fig. 3 Schematic diagram depicting the genomisotopic approach for isolation of Orfamide A from cultures of Pf-5. (**a**) The orfamide A biosynthesis genes, *ofaA*, *ofaB*, and *ofaC*, which encode a non-ribosomal peptide synthetase (NRPS) (**b**) Ten modules (M1–M10) were identified in the deduced protein sequence of *ofaA*, *ofaB*, and *ofaC*, each with a condensation (C), adenylation (A), and thiolation (T) domain. Two thioesterase (TE) domains were identified in the 3′ end of OfaC. Amino acids incorporated into the final peptide product were predicted by bioinformatic analysis of the A domain of each module. The specified amino acid is shown above the corresponding A domain. (**c**) Pf-5 was grown under culture conditions favourable to the expression of *ofaA*, *ofaB*, and *ofaC*, as determined using RT-PCR. Isotopically-labelled leucine, an amino acid predicted to be present in the final peptide product, was added to the culture medium and the label was used to guide a fractionation scheme for purification of the natural product. (**d**) The structure of orfamide A, which was isolated using the genomisotopic approach. Adapted from Gross et al. (2007)

length of the peptidyl and lipid moieties: the viscosin, amphisin, tolaasin syringomycin, and putisolvin groups (Raaijmakers et al. 2006). The size of the peptide portion ranges from nine amino acids (viscosin group) up to 25 amino acids (syringopeptin group), but there are no previously described CLPs from *Pseudomonas* spp. with a 10 amino acid peptide chain.

The genomisotopic approach

With the increasing number of genomes sequenced and available in the public domain, a large number of orphan gene clusters, for which the encoded natural product is unknown, have been identified. These orphan gene clusters represent a tremendous source of novel and possibly bioactive compounds, and efficient methods to exploit this valuable resource are needed. Recently, a novel approach termed the genomisotopic (GI) approach was developed to meet that need, with the CLP produced by Pf-5 selected as the first orphan metabolite to be isolated using this approach (Gross et al. 2007). The GI approach uses an isotopically-labelled amino acid, predicted to be a precursor of a peptide from bioinformatic analysis of the NRPS nucleotide sequence, to guide fractionation for the purification of a final peptide product (Fig. 3c). Using a two-pronged approach employing GI and traditional assay-guided fractionation in parallel, novel CLPs termed orfamides were identified from cultures of Pf-5 (Gross et al. 2007). The chemical structure of these molecules conforms to the amino acid sequence predicted bioinformatically from the sequence of A domains of OfaA, OfaB, and OfaC.

Characteristics of the orfamides

The orfamides are founding members of a new class of CLPs characterized by a 3-hydroxy dodecanoic or tetradecanoic (myristic) acid connected to the N-terminus of a 10 amino acid cyclic peptide. Orfamide A (Fig. 3d) is the dominant CLP produced by Pf-5. The amino and fatty acid partial structures of orfamide A were determined by standard amino acid analysis, HR-ESI-TOF-MS and an extensive array of 1D and 2D NMR experiments. The sequence of the residues was accomplished using a combination of MS/MS as well as HMBC and ROESY NMR experiments. Chiral GC-MS and a modified Marfey's analysis were used to establish the absolute configurations of each amino acid (Gross et al. 2007).

Although orfamide A is the dominant CLP produced by Pf-5, related compounds (also called orfamides) are produced in much smaller amounts by the bacterium (Gross et al. 2007). Orfamide B differs from orfamide A by a substitution of valine for *D*-allo isoleucine at the fourth position on the peptide chain. Orfamide C differs from orfamide A by the substitution of dodecanoic acid for tetradecanoic acid as the lipid moiety. A mutation in the NRPS-encoding gene *ofaB* eliminates production of all three orfamides by Pf-5 (Gross et al. 2007), indicating that all three compounds are products of the same gene cluster. The relaxed substrate specificity of the A-domain responsible for the activation of L-Ile could explain the production of slight amounts of orfamide B by Pf-5. Similarly, the relaxed substrate specificity of the C domain of module 1 for tetradecanoic acid could explain the altered fatty acid present in orfamide C. In addition to the three characterized orfamides, other compounds with a similar polarity and mass range, which are present in trace amounts in culture supernatants of Pf-5, were not detected in cultures of the *ofaB* mutant. It is likely that a range of orfamides are produced in very small concentrations by Pf-5, each of which reflects a relaxed specificity of one or more of the domains in the NRPS responsible for orfamide biosynthesis. Small amounts of derivatives of viscosins or amphisins are produced by other strains of *P. fluorescens*, which has also been attributed to the relaxed specificities of biosynthetic enzymes (De Souza et al. 2003; Nielsen et al. 2002).

The orfamide biosynthetic gene cluster

The orfamide biosynthesis gene cluster (Fig. 2) includes two putative regulatory genes that encode transcriptional regulators in the LuxR family: PFL_2143 and PFL_2150. Three genes (PFL_2142, PFL_2148, and PFL_2149) with predicted roles in transport have also been described. The predicted peptide sequences of these genes have domains characteristic of the components of an ABC transport system: an outer membrane porin (PFL_2142), a cytoplasmic membrane protein with ATP binding domains (PFL2149), and a membrane fusion protein (PFL_2148). The organization of the nine genes in the proposed orfamide gene cluster (Fig. 2) is identical to that of a CLP biosynthetic gene cluster in *P. fluorescens* Pf0-1 (Accession no. NC_007492). The predicted amino acid sequences of the genes in the orfamide gene cluster are also very similar (ranging from 62% to 87% identical) to homologs in the Pf0-1 genome. Therefore, it appears that the CLP biosynthesis gene cluster is conserved among the two strains.

Genes for the biosynthesis of the lipid side chain of orfamide have not been identified. Enzymes for fatty acid biosynthesis, such as β-ketoacyl-ACP synthetases or acyl-CoA ligases, are not present in the orfamide gene cluster. Also absent from the gene cluster is an acetyltransferase, purportedly required for the acyltransfer of a 3-hydroxy fatty acid to the first amino acid of the peptide chain. While biosynthetic gene clusters for CLPs can include genes for synthesis of the lipid moiety, as for iturin biosynthesis by *Bacillus subtilis* (Tsuge et al. 2001), this is not always the case. For example, the large genomic island in *P. syringae* B728a, which contains the NRPSs for biosynthesis of the CLPs syringomycin and syringopeptin, lacks genes for fatty acid biosynthesis (Feil et al. 2005). Therefore, as in Pf-5, genes for the biosynthesis of the lipid side chains are not co-located with the NRPSs for biosynthesis of these CLPs in the genome of *P. syringae* B728a. It is possible that the lipid moiety is provided from primary metabolism.

Biological significance of the orfamides

Due in part to their biosurfactant properties, CLPs influence the lifestyle of *Pseudomonas* in many ways, facilitating bacterial movement and growth on agar or seed surfaces (Nielsen et al. 2005), influencing surface adhesion and altering biofilm development and stability (Raaijmakers et al. 2006). The amphiphilic properties of CLPs influence the motility of bacterial cells on surfaces, and mutants deficient in CLP production commonly exhibit diminished swarming motility (Kuiper et al. 2004; Roongsawang et al. 2003). Accordingly, an orfamide A deficient mutant of Pf-5 exhibits reduced swarming motility compared to the parental strain (Gross et al. 2007). The surfactant properties of CLPs are also likely to influence adhesion of bacterial cells to surfaces, an important component of biofilm formation. Nevertheless, different CLPs have different reported effects on adhesion and biofilm formation by *Pseudomonas* spp. Biofilm formation was enhanced in a mutant of *P. putida* PCL1445 deficient in the production of the CLP putisolvin. Furthermore, purified putisolvin I broke down established biofilms produced by PCL1445 (Kuiper et al. 2004). A mutant of *Pseudomonas* sp. MIS38 deficient in the production of the CLP arthrofactin produced less stable but more abundant biofilms than the parental strain (Roongsawang et al. 2003). CLP deficient mutants of *P. fluorescens* SWB25 exhibited diminished biofilm formation (de Bruijn et al. 2007). An orfamide-deficient mutant did not differ statistically from Pf-5 in adhesion to polystyrene, a key assay indicative of biofilm formation (Gross et al. 2007). Due perhaps to the diversity of the CLPs and CLP-producing strains, there appears to be no uniform role of these molecules in surface adhesion by *Pseudomonas* spp.

The structural diversity of CLPs is reflected in the diverse biological activities of this class of molecules, members of which act as phytotoxins or anti-microbial agents (Raaijmakers et al. 2006; Nybroe and Sørensen 2004). Certain CLPs isolated from biocontrol strains of *Pseudomonas* spp. inhibit mycelial growth or development of fungal plant pathogens, including the soilborne fungus *Rhizoctonia solani* (Miller et al. 1998; Nielsen et al. 2000, 2002). In contrast, purified orfamide A did not inhibit mycelial growth of *R. solani* on agar surfaces (Gross et al. 2007). Zoospores of Oomycetes, lacking a cell wall, are vulnerable to membrane disruption caused by CLPs (de Souza et al. 2003), effected either through the non-specific detergent properties of CLPs, their capacity to form transmembrane pores (Nybroe and Sørensen 2004), or other mechanisms. Consequently, zoospores of *Phytophthora* spp. lyse within seconds of exposure to CLPs such as a viscosin-like CLP produced by SBW25 (de Bruijn et al. 2007) or orfamide A (Gross et al. 2007). In contrast, the CLP viscosinamide decreased zoospore production and increased zoospore encystment by another Oomycete, *Pythium ultimum* (Thrane et al. 2000). As the only motile propagule of the Oomycetes, zoospores represent a critical stage in the disease cycle, especially for plants grown hydroponically or in water-saturated soils. Surfactants, produced synthetically or by biological control agents, can provide effective management of diseases caused by Oomycetes on plants grown hydroponically (Stanghellini and Miller 1997). Accordingly, a CLP related to massetolide A, which lyses zoospores of several Oomycete pathogens, is also thought to play a critical role in biological control of root rot caused by *P. ultimum* (de Souza et al. 2003). With the discovery of the orfamides, knowledge of the structural diversity of the CLPs has been expanded, as well as their potential effects as factors influencing the interactions of rhizosphere bacteria with other components of the soil microflora and the plant host.

Conclusions

The genomic sequence of the rhizosphere bacterium *P. fluorescens* Pf-5 provides a variety of insights into this organism's lifestyle in association with plants in the natural environment. It revealed pathways for utilization of an extensive array of carbon substrates found in plant root and seed exudates as well as siderophores produced by diverse soil inhabiting microorganisms. The expanded transport and oxidative stress-resistance capacities found in the genome are likely to provide a foundation for the fitness of *P. fluorescens* Pf-5 in the rhizosphere. At 7.07 Mb, the genome of *P. fluorescens* Pf-5 is larger than those of other *Pseudomonas* spp. sequenced to date, a difference that can be attributed in part to the presence of at least nine gene clusters for secondary metabolite production. Four of these gene clusters specify the biosynthesis of antibiotics with a well established role in biological control, and the organization and sequences of the biosynthetic, transport, and regulatory genes therein are very similar to those described in other strains of *Pseudomonas* spp. The genomic sequence also highlighted three orphan gene clusters, which encode for natural products with unknown structures and biological activities. The structure of one of these orphan metabolites could be predicted bioinformatically from the sequence of a NRPS within the gene cluster. The bioinformatically-predicted structure has since been confirmed, and the product defined as orfamide A, the founding member of a new class of CLPs. Like other CLPs produced by *Pseudomonas* spp., orfamide A functions as a biosurfactant, influencing swarming motility of Pf-5 and lysing zoospores produced by an Oomycete plant pathogen. Recently, the structure of a viscosin-like CLP produced by *P. fluorescens* strain SBW25 was reported, and shown to conform to the structure predicted bioinformatically from the genomic sequence of strain SBW25. While the path from genomic sequence data to chemical discovery has been lightly trod to date, the discovery of orfamide A provides one of the first examples of genome mining from *Pseudomonas* spp. *Pseudomonas* spp. are prolific producers of antibiotics, and the availability of genomic sequences for several *Pseudomonas* spp. now opens the door for discovery of novel natural products with potential roles in the ecology and plant growth-promoting properties of these bacteria.

Acknowledgements The Pf-5 genomic sequencing project was supported by Initiative for Future Agriculture and Food Systems Grant no. 2001-52100-11329 from the U.S. Department of Agriculture Cooperative State Research, Education, and Extension Service. Research in the Loper laboratory is funded by the U.S. Department of Agriculture, Agricultural Research Service CRIS project 5358-12220-002-00D. H.G. thanks the German Research Foundation for a Research Fellowship (GR 2673/1-1).

References

Abbas, A., McGuire, J. E., Crowley, D., Baysse, C., Dow, M., & O'Gara, F. (2004). The putative permease PhlE of *Pseudomonas fluorescens* F113 has a role in 2,4-diacetylphloroglucinol resistance and in general stress tolerance. *Microbiology, 150*, 2443–2450.

Abbas, A., Morrissey, J. P., Marquez, P. C., Sheehan, M. M., Delany, I. R., & O'Gara, F. (2002). Characterization of interactions between the transcriptional repressor PhlF and its binding site at the *phlA* promoter in *Pseudomonas fluorescens* F113. *Journal of Bacteriology, 184*, 3008–3016.

Ansari, M. Z., Yadav, G., Gokhale, R. S., & Mohanty, D. (2004). NRPS-PKS: A knowledge-based resource for analysis of NRPS/PKS megasynthases. *Nucleic Acids Research, 32*, W405–W413.

Baehler, E., Bottiglieri, M., Pechy-Tarr, M., Maurhofer, M., & Keel, C. (2005). Use of green fluorescent protein-based reporters to monitor balanced production of antifungal compounds in the biocontrol agent *Pseudomonas fluorescens* CHA0. *Journal of Applied Microbiology, 99*, 24–38.

Bangera, M. G., & Thomashow, L. S. (1999). Identification and characterization of a gene cluster for synthesis of the polyketide antibiotic 2,4-diacetylphloroglucinol from *Pseudomonas fluorescens* Q2-87. *Journal of Bacteriology, 181*, 3155–3163.

Blumer, C., & Haas, D. (2000). Mechanism, regulation, and ecological role of bacterial cyanide biosynthesis. *Archives of Microbiology, 173*, 170–177.

Bottiglieri, M., & Keel, C. (2006). Characterization of PhlG, a hydrolase that specifically degrades the antifungal compound 2,4-diacetylphloroglucinol in the biocontrol agent *Pseudomonas fluorescens* CHA0. *Applied and Environmental Microbiology, 72*, 418–427.

Brodhagen, M., Paulsen, I., & Loper, J. E. (2005). Reciprocal regulation of pyoluteorin production with membrane transporter gene expression in *Pseudomonas fluorescens* Pf-5. *Applied and Environmental Microbiology, 71*, 6900–6909.

Challis, G. L., Ravel, J., & Townsend, C. A. (2000). Predictive, structure-based model of amino acid recognition by nonribosomal peptide synthetase adenylation domains. *Chemistry & Biology, 7*, 211–224.

Cipollone, R., Frangipani, E., Tiburzi, F., Imperi, F., Ascenzi, P., & Visca, P. (2007). Involvement of *Pseudomonas aeruginosa* rhodanese in protection from cyanide toxicity. *Applied and Environmental Microbiology, 73*, 390–398.

Cooper, M., Tavankar, G. R., & Williams, H. D. (2003). Regulation of expression of the cyanide-insensitive terminal oxidase in *Pseudomonas aeruginosa*. *Microbiology, 149*, 1275–1284.

Cronin D., Moënne-Loccoz, Y., Fenton, A., Dunne, C., Dowling, D. N., & O'Gara, F. (1997). Role of 2,4-diacetylphloroglucinol in the interactions of the biocontrol pseudomonad strain F113 with the potato cyst nematode *Globodera rostochiensis*. *Applied and Environmental Microbiology, 63*, 1357–1361.

Daborn, P. J., Waterfield, N., Silva, C. P., Au, C. P. Y., Sharma, S., & ffrench-Constant, R. H. (2002). A single *Photorhabdus* gene, makes caterpillars floppy (*mcf*), allows *Escherichia coli* to persist within and kill insects. *Proceedings of the National Academy of Sciences of the United States of America, 99*, 10742–10747.

Delany, I., Sheehan, M. M., Fenton, A., Bardin, S., Aarons, S., & O'Gara, F. (2000). Regulation of production of the antifungal metabolite 2,4-diacetylphloroglucinol in *Pseudomonas fluorescens* F113: genetic analysis of *phl*F as a transcriptional repressor. *Microbiology, 146*, 537–543.

de Bruijn, I., de Kock, M. J. D., Yang, M., de Waard, P., van Beek, T. A., & Raaijmakers, J. M. (2007). Genome-based discovery, structure prediction and functional analysis of cyclic lipopeptide antibiotics in *Pseudomonas* species. *Molecular Microbiology, 63*, 417–428.

De Souza, J. T., De Boer, M., De Waard, P., Van Beek, T. A., & Raaijmakers, J. M. (2003). Biochemical, genetic, and zoosporicidal properties of cyclic lipopeptide surfactants produced by *Pseudomonas fluorescens*. *Applied and Environmental Microbiology, 69*, 7161–7172.

Dong, C., Flecks, S., Unversucht, S., Haupt, C., van Pée, K.-H., & Naismith, J. H. (2005). Tryptophan 7-halogenase (PrnA) structure suggests a mechanism for regioselective chlorination. *Science, 309*, 2216–2219.

dos Santos, V. A. P. M., Heim, S., Moore, E. R. B., Strätz, M., & Timmis, K. N. (2004). Insights into the genomic basis of niche specificity of *Pseudomonas putida* KT2440. *Environmental Microbiology, 6*, 1264–1286.

Feil, H., Feil, W. S., Chain, P., Larimer, F., DiBartolo, G., Copeland, A., Lykidis, A., Trong, S., Nolan, M., Goltsman, E., Thiel, J., Malfatti, S., Loper, J. E., Lapidus, A., Detter, J. C., Land, M., Richardson, P. M., Kyrpides, N. C., Ivanova, N., & Lindow, S. E. (2005). Comparison of the complete genome sequences of *Pseudomonas syringae* pv. *syringae* B728a and pv. *tomato* DC3000. *Proceedings of the National Academy of Sciences of the United States of America, 102*, 11064–11069.

Finking, R., & Marahiel, M. A. (2004). Biosynthesis of nonribosomal peptides. *Annual Review of Microbiology, 58*, 453–488.

Folders, J., Algra, J., Roelofs, M. S., van Loon, L. C., Tommassen, J., & Bitter, W. (2001). Characterization of *Pseudomonas aeruginosa* chitinase, a gradually secreted protein. *Journal of Bacteriology, 183*, 7044–7052.

Gross, H., Stockwell, V. O., Henkels, M. D., Nowak-Thompson, B., Loper, J. E., & Gerwick, W. H. (2007). The genomisotopic approach: a systematic method to isolate products of orphan biosynthetic gene clusters. *Chemistry & Biology, 14*, 53–63.

Guenzi, E., Galli, G., Grgurina, I., Gross, D. C., & Grandi, G. (1998). Characterization of the syringomycin synthetase gene cluster. A link between prokaryotic and eukaryotic peptide synthetases. *Journal of Biological Chemistry, 273*, 32857–32863.

Haas, D., & Defago, G. (2005). Biological control of soil-borne pathogens by fluorescent pseudomonads. *Nature Reviews Microbiology, 3*, 307–319.

Haas, D., & Keel, C. (2003). Regulation of antibiotic production in root-colonizing *Pseudomonas* spp. and relevance for biological control of plant disease. *Annual Review of Phytopathology, 41*, 117–153.

Hammer, P. E., Burd, W., Hill, D. S., Ligon, J. M., & van Pée, K. (1999). Conservation of the pyrrolnitrin biosynthetic gene cluster among six pyrrolnitrin-producing strains. *FEMS Microbiology Letters, 180*, 39–44.

Hammer, P. E., Hill, D. S., Lam, S. T., Van Pée, K. H., & Ligon, J. M. (1997). Four genes from *Pseudomonas fluorescens* that encode the biosynthesis of pyrrolnitrin. *Applied and Environmental Microbiology, 63*, 2147–2154.

Harman, G. E., Howell, C. R., Viterbo, A., Chet, I., & Lorito, M. (2004). *Trichoderma* species – opportunistic, avirulent plant symbionts. *Nature Reviews Microbiology, 2*, 43–56.

Helmann, J. D. (2002). The extracytoplasmic function (ECF) sigma factors. *Advances in Microbial Physiology, 46*, 47–110.

Howell, C. R., & Stipanovic, R. D. (1979). Control of *Rhizoctonia solani* in cotton seedlings with *Pseudomonas fluorescens* and with an antibiotic produced by the bacterium. *Phytopathology, 69*, 480–482.

Howell, C. R., & Stipanovic, R. D. (1980). Suppression of *Pythium ultimum* induced damping-off of cotton seedlings by *Pseudomonas fluorescens* and its antibiotic pyoluteorin. *Phytopathology, 70*, 712–715.

Huang, X. Q., Yan, A., Zhang, X. H., & Xu, Y. Q. (2006). Identification and characterization of a putative ABC transporter PltHIJKN required for pyoluteorin production in *Pseudomonas* sp. M18. *Gene, 376*, 68–78.

Huang, X., Zhu, D., Ge, Y., Hu, H., Zhang, X., & Xu, Y. (2004). Identification and characterization of *pltZ*, a gene involved in the repression of pyoluteorin biosynthesis in *Pseudomonas* sp. M18. *FEMS Microbiology Letters, 232*, 197–202.

Keel, C., Schnider, U., Maurhofer, M., Voisard, C., Laville, J., Burger, U., Wirthner, P., Haas, D., & Défago, G. (1992). Suppression of root diseases by *Pseudomonas fluorescens* CHA0: importance of the bacterial secondary metabolite 2,4-diacetylphloroglucinol. *Molecular Plant-Microbe Interactions, 5*, 4–13.

Kiil, K., Binnewies, T. T., Sicheritz-Pontén, T., Willenbrock, H., Hallin, P. F., Wassenaar, P. F., & Ussery, D. W. (2005a). Genome update: sigma factors in 240 bacterial genomes. *Microbiology, 151*, 3147–3150.

Kiil, K., Ferchaud, J. P., David, C., Binnewies, T. T., Wu, H., Sicheritz-Pontén, T., Willenbrock, H., & Ussery, D. W. (2005b). Genome update: distribution of two-component transduction systems in 250 bacterial genomes. *Microbiology, 151*, 3447–3452.

Kirner, S., Hammer, P. E., Hill, D. S., Altmann, A., Fischer, I., Weislo, L. J., Lanahan, M., van Pée, K. H., & Ligon, J. M. (1998). Functions encoded by pyrrolnitrin biosynthetic genes from *Pseudomonas fluorescens*. *Journal of Bacteriology, 180*, 1939–1943.

Knowles, C. J. (1976). Microorganisms and cyanide. *Bacteriological Reviews, 40*, 652–680.

Kraus, J., & Loper, J. E. (1992). Lack of evidence for a role of antifungal metabolite production by *Pseudomonas fluorescens* Pf-5 in biological control of Pythium damping-off of cucumber. *Phytopathology, 82*, 264–271.

Kraus, J., & Loper, J. E. (1995). Characterization of a genomic region required for production of the antibiotic pyoluteorin by the biological control agent *Pseudomonas fluorescens* Pf-5. *Applied and Environmental Microbiology, 61*, 849–854.

Kuiper, I., Lagendijk, E. L., Pickford, R., Derrick, J. P., Lamers, G. E. M., Thomas-Oates, J. E., Lugtenberg, B. J. J., & Bloemberg, G. V. (2004). Characterization of two *Pseudomonas putida* lipopeptide biosurfactants, putisolvin I and II, which inhibit biofilm formation and break down existing biofilms. *Molecular Microbiology, 51*, 97–113.

Laville, J., Blumer, C., Von Schroetter, C., Gaia, V., Defago, G., Keel, C., & Haas, D. (1998). Characterization of the *hcnABC* gene cluster encoding hydrogen cyanide synthase and anaerobic regulation by ANR in the strictly aerobic biocontrol agent *Pseudomonas fluorescens* CHA0. *Journal of Bacteriology, 180*, 3187–3196.

Ligon, J. M., Hill, D. S, Hammer, P. E., Torkewitz, N. R., Hofmann, D., Kempf, H.-J., & van Pée, K.-H. (2000). Natural products with antifungal activity from *Pseudomonas* biocontrol bacteria. *Pest Management Science, 56*, 688–695.

Loper, J. E., & Henkels, M. D. (1999). Utilization of heterologous siderophores enhances levels of iron available to *Pseudomonas putida* in the rhizosphere. *Applied and Environmental Microbiology, 65*, 5357–5363.

Loper, J. E., Kobayashi, D. Y., & Paulsen, I. T. (2007). The genomic sequence of *Pseudomonas fluorescens* Pf-5: insights into biological control. *Phytopathology, 97*, 233–238.

Maurhofer, M., Keel, C., Haas, D., & Défago, G. (1995). Influence of plant species on disease suppression by *Pseudomonas fluorescens* strain CHA0 with enhanced antibiotic production. *Plant Pathology, 44*, 40–50.

Mavrodi, D. V., Blankenfeldt, W., & Thomashow, L. S. (2006). Phenazine compounds in fluorescent *Pseudomonas* spp. biosynthesis and regulation. *Annual Review of Phytopathology, 44*, 417–445.

Mavrodi, D. V., Paulsen, I. T., Ren, Q., & Loper, J. E. (2007). Genomics of *Pseudomonas fluorescens* Pf-5. In A. Filloux & J. L. Ramos (Eds.), *Pseudomonas* (Vol. V, pp. 1–28). The Netherlands: Springer Press.

Mazurier, S., Lemunier, M., Siblot, S., Mougel, C., & Lemanceau, P. (2004). Distribution and diversity of type III secretion system-like genes in saprophytic and phytopathogenic fluorescent pseudomonads. *FEMS Microbiology Ecology, 49*, 455–467.

Mellano, M. A., & Cooksey, D. A. (1998). Nucleotide sequence and organization of copper resistance genes from *Pseudomonas syringae* pv. *tomato*. *Journal of Bacteriology, 170*, 2879–2883.

Miller, C. M., Miller, R. V., Garton-Kenny, D., Redgrave, B., Sears, J., Condron, M. M., Teplow, D. B., & Strobel, G. A. (1998). Ecomycins, unique antimycotics from *Pseudomonas viridiflava*. *Journal of Applied Microbiology, 84*, 937–944.

Nielsen, T. H., Nybroe, O., Koch, B., Hansen, M., & Sorensen, J. (2005). Genes involved in cyclic lipopeptide production are important for seed and straw colonization by *Pseudomonas* sp. strain DSS73. *Applied and Environmental Microbiology, 71*, 4112–4116.

Nielsen, T. H., Sorensen, D., Tobiasen, C., Andersen, J. B., Christophersen, C., Givskov, M., & Sorensen, J. (2002). Antibiotic and biosurfactant properties of cyclic lipopeptides produced by fluorescent *Pseudomonas* spp. from the sugar beet rhizosphere. *Applied and Environmental Microbiology, 68*, 3416–3423.

Nielsen, T. H., Thrane, C., Christophersen, C., Anthoni, U., & Sorensen, J. (2000). Structure, production characteristics and fungal antagonism of tensin – a new antifungal cyclic lipopeptide from *Pseudomonas fluorescens* strain 96.578. *Journal of Applied Microbiology, 89*, 992–1001.

Nowak-Thompson, B., Chaney, N., Wing, J. S., Gould, S. J., & Loper, J. E. (1999). Characterization of the pyoluteorin biosynthetic gene cluster of *Pseudomonas fluorescens* Pf-5. *Journal of Bacteriology, 181*, 2166–2174.

Nowak-Thompson, B., Gould, S. J., Kraus, J., & Loper, J. E. (1994). Production of 2,4-diacetylphloroglucinol by the biocontrol agent *Pseudomonas fluorescens* Pf-5. *Canadian Journal of Microbiology, 40*, 1064–1066.

Nowak-Thompson, B., Gould, S. J., & Loper, J. E. (1997). Identification and sequence analysis of the genes encoding a polyketide synthase required for pyoluteorin biosynthesis in *Pseudomonas fluorescens* Pf-5. *Gene, 204*, 17–24.

Nybroe, O., & Sørensen, J. (2004). Production of cyclic lipopeptides by fluorescent pseudomonads. In J. L. Ramos (Ed.), *Pseudomonas, Vol. 3, Biosynthesis of Macromolecules and Molecular Metabolism* (pp. 147–172). New York: Kluwer Academic/Plenum Publishers.

Parret, A. H. A., Temmerman, K., & De Mot, R. (2005). Novel lectin-like bacteriocins of biocontrol strain *Pseudomonas fluorescens* Pf-5. *Applied and Environmental Microbiology, 71*, 5197–5207.

Paulsen, I. T., Press, C. M., Ravel, J., Kobayashi, D. Y., Myers, G. S., Mavrodi, D. V., DeBoy, R. T., Seshadri, R., Ren, Q., Madupu, R., Dodson, R. J., Durkin, A. S., Brinkac, L. M., Daugherty, S. C., Sullivan, S. A., Rosovitz, M. J., Gwinn, M. L., Zhou, L., Schneider, D. J., Cartinhour, S. W., Nelson, W. C., Weidman, J., Watkins, K., Tran, K., Khouri, H., Pierson, E. A., Pierson, L. S. III, Thomashow, L. S., & Loper, J. E. (2005). Complete genome sequence of the plant commensal *Pseudomonas fluorescens* Pf-5. *Nature Biotechnology, 23*, 873–878.

Pfender, W. F., Kraus, J., & Loper, J. E. (1993). A genomic region from *Pseudomonas fluorescens* Pf-5 required for pyrrolnitrin production and inhibition of *Pyrenophora tritici-repentis* in wheat straw. *Phytopathology, 83*, 1223–1228.

Preston, G. M., Bertrand, N., & Rainey, P. B. (2001). Type III secretion in plant growth-promoting *Pseudomonas fluorescens* SBW25. *Molecular Microbiology, 41*, 999–1014.

Raaijmakers, J. M., de Bruijn, I., & de Kock, M. J. (2006). Cyclic lipopeptide production by plant-associated *Pseudomonas* spp.: diversity, activity, biosynthesis, and regulation. *Molecular Plant-Microbe Interactions, 19*, 699–710.

Raaijmakers, J. M., Vlami, M., & de Souza, J. T. (2002). Antibiotic production by bacterial biocontrol agents. *Antonie van Leeuwenhoek, 81*, 537–547.

Ramette, A., Frapolli, M., Défago, G., & Moënne-Loccoz, Y. (2003). Phylogeny of HCN synthase-encoding *hcnBC* genes in biocontrol fluorescent pseudomonads and its relationship with host plant species and HCN synthesis ability. *Molecular Plant-Microbe Interactions, 16*, 525–535.

Rao, K. V., & Reddy, G. C. (1990). Synthesis and herbicidal activity of the halo analogues of pyoluteorin. *Journal of Agricultural and Food Chemistry, 38*, 1260–1263.

Rezzonico, F., Binder, C., Defago, G., & Moenne-Loccoz, Y. (2005). The type III secretion system of biocontrol *Pseudomonas fluorescens* KD targets the phytopathogenic chromista *Pythium ultimum* and promotes cucumber protection. *Molecular Plant-Microbe Interactions, 18*, 991–1001.

Rodriguez, F., & Pfender, W. F. (1997). Antibiosis and antagonism of *Sclerotinia homoeocarpa* and *Drechslera poae* by *Pseudomonas fluorescens* Pf-5 *in vitro* and *in planta*. *Phytopathology, 87*, 614–621.

Roongsawang, N., Hase, K., Haruki, M., Imanaka, T., Morikawa, M., & Kanaya, S. (2003). Cloning and characterization of the gene cluster encoding arthrofactin synthetase from *Pseudomonas* sp. MIS38. *Chemistry & Biology, 10*, 869–880.

Schnider-Keel, U., Seematter, A., Maurhofer, M., Blumer, C., Duffy, B., Gigot-Bonnefoy, C., Reimmann, C., Notz, R., Defago, G., Haas, D., & Keel, C. (2000). Autoinduction of 2,4-diacetylphloroglucinol biosynthesis in the biocontrol agent *Pseudomonas fluorescens* CHA0 and repression by the bacterial metabolites salicylate and pyoluteorin. *Journal of Bacteriology, 182*, 1215–1225.

Scholz-Schroeder, B. K., Soule, J. D., & Gross, D. C. (2003). The *sypA*, *sypS*, and *sypC* synthetase genes encode twenty-two modules involved in the nonribosomal peptide synthesis of syringopeptin by *Pseudomonas syringae* pv. *syringae* B301D. *Molecular Plant-Microbe Interactions, 16*, 271–280.

Sharifi-Tehrani, A., Zala, M., Natsch, A., Moenne-Loccoz, Y., & Defago, G., (1998). Biocontrol of soil-borne fungal plant diseases by 2,4-diacetylphloroglucinol-producing fluorescent pseudomonads with different restriction profiles of amplified 16S rDNA. *European Journal of Plant Pathology, 104*, 631–643.

Siddiqui, I. A., Haas, D., & Heeb, S. (2005). Extracellular protease of *Pseudomonas fluorescens* CHA0, a biocontrol factor with activity against the root-knot nematode *Meloidogyne incognita*. *Applied and Environmental Microbiology, 71*, 5646–5649.

Stachelhaus, T., Mootz, H. D., & Marahiel, M. A. (1999). The specificity-conferring code of adenylation domains in nonribosomal peptide synthetases. *Chemistry & Biology, 6*, 493–505.

Stanghellini, M. E., & Miller, R. M. (1997). Biosurfactants: their identity and potential efficacy in the biological control of zoosporic plant pathogens. *Plant Disease, 81*, 4–12.

Stutz, E. W., Defago, G., & Kern, H. (1986). Naturally occurring fluorescent pseudomonads involved in suppression of black root rot of tobacco. *Phytopathology, 76*, 181–185.

Thrane, C., Nielsen, T. H., Nielsen, M. N., Sorensen, J., & Olsson, S. (2000). Viscosinamide-*Pseudomonas fluorescens* DR54 exerts a biocontrol effect on *Pythium ultimum* in sugar beet rhizosphere. *FEMS Microbiology Ecology, 33*, 139–146.

Tripathi, R. K., & Gottlieb, D. (1969). Mechanism of action of the antifungal antibiotic pyrrolnitrin. *Journal of Bacteriology, 100*, 310–318.

Tsuge, K., Akiyama, T., & Shoda, M. (2001). Cloning, sequencing, and characterization of the iturin A operon. *Journal of Bacteriology, 183*, 6265–6273.

Voisard, C., Keel, C., Haas., D., & Défago, G. (1989). Cyanide production by *Pseudomonas fluorescens* helps suppress black root rot of tobacco under gnotobiotic conditions. *EMBO Journal, 8*, 351–358.

Weller, D. M., Landa, B. B., Mavrodi, O. V., Schroeder, K. L., De La Fuente, L., Blouin Bankhead, S., Allende Molar, R., Bonsall, R. F., Mavrodi, D. V., & Thomashow, L. S. (2007). Role of 2,4-diacetylphloroglucinol-producing fluorescent *Pseudomonas* spp. in the defense of plant roots. *Plant Biology, 9*, 4–20.

Weller, D. M., Raaijmakers, J. M., McSpadden Gardener, B. B., & Thomashow, L. S. (2002). Microbial populations responsible for specific soil suppressiveness to plant pathogens. *Annual Review of Phytopathology, 40*, 309–348.

Wenzel, S. C., & Müller, R. (2005). Formation of novel secondary metabolites by bacterial multimodular assembly lines: Deviations from textbook biosynthetic logic. *Current Opinion in Chemical Biology, 9*, 447–458.

Xu, G. W., & Gross, D. C. (1986). Selection of fluorescent pseudomonads antagonistic to *Erwinia carotovora* and suppressive of potato seed piece decay. *Phytopathology, 76*, 414–422.

Eur J Plant Pathol (2007) 119:279–300
DOI 10.1007/s10658-007-9186-9

REVIEW

The magic and menace of metagenomics: prospects for the study of plant growth-promoting rhizobacteria

Johan H. J. Leveau

Received: 25 December 2006 / Accepted: 9 July 2007 / Published online: 21 September 2007

Abstract This article aims to be a pragmatic primer into the field of metagenomics with special emphasis on the prospective contributions of metagenomics to the study of plant growth-promoting rhizobacteria (PGPR). After an introduction into the concepts and methodologies of metagenomics and a discussion of the numerous emerging variations on the basic theme, there will follow a short overview of the success stories in metagenomics (the 'magic' in the title of this review), a brief discussion about the technical problems and unrealistic expectations that are sometimes associated with metagenomics (the 'menace'), and a shortlist of the lessons that can be learned by those that wish to explore the utility of metagenomics in the study of PGPR.

Keywords Metagenomics · PGPR · Plant growth-promoting rhizobacteria · Environmental shotgun sequencing

Metagenomics

The phrase 'metagenome of the soil' was first used by Handelsman et al. (1998) to describe the collective genomes of soil microflora. In the area of microbial ecology, the term 'metagenomics' is now synonymous with the culture-independent application of genomics techniques to the study of microbial communities in their natural environments (Chen and Pachter 2005). Metagenomics arose in reaction to the observation that the majority of microorganisms on Earth resist life in captivity, i.e. they cannot be grown in broth or on plates in the laboratory. An often-cited estimate is that as much as 99% or more of microbial life remains unculturable, and therefore cannot be studied and understood in a way that microbial ecologists have become accustomed to over the past century. Metagenomics exploits the fact that while some microorganisms are culturable and others are not, all of them (i.e., 100%) are life-forms based on DNA as a carrier of genetic information. The metagenomic toolbox allows accessing, storing, and analyzing this DNA and thus can provide an otherwise hard-to-attain insight into the biology and evolution of environmental microorganisms, independent of their culturable status.

In modern metagenomics, three major and oftentimes overlapping directions can be recognized. The first is aimed at linking phylogeny to function. Once microbial ecologists got a satisfactory grasp on the issue of 'who is out there?' (Amann 2000), they set out to answer the question 'who's doing what out there?'. As one of many complementary methodologies, metagenomics can help answer that question in an indirect, culture-independent manner. One example

J. H. J. Leveau (✉)
Netherlands Institute of Ecology (NIOO-KNAW),
Heteren, The Netherlands
e-mail: j.leveau@nioo.knaw.nl

is through phylogenetic anchoring (Riesenfeld et al. 2004b), which involves the screening of large-insert environmental libraries for clones that carry phylogenetically informative genes (to reveal the 'who') and analyzing their flanking DNA for genes that reveal possible environmental functions of the DNA's owner. A second trend in modern metagenomics involves its exploitation for the discovery of enzymes with novel, industrial and possibly exploitable properties. This aspect of metagenomics was well-reviewed recently by Lorenz and Eck (2005), who conclude that metagenomics "provides industry with an unprecedented chance to bring biomolecules into industrial application". The third and most recent trend in metagenomics is the mass sequencing of environmental samples. The promise of this approach is to offer a more global or systems-biology view of the community under study. Indeed, in several instances has mass sequencing led to more complete assessments of genetic diversity and to first insights into the interactivities that occur in microbial communities (DeLong et al. 2006; Edwards et al. 2006; Gill et al. 2006; Hallam et al. 2004; Schmeisser et al. 2003; Tringe et al. 2005; Tyson et al. 2004; Venter et al. 2004).

Early pioneers in the field of metagenomics were Schmidt et al. (1991) who studied the phylogenetic diversity of an oligotrophic marine picoplankton community in the north central Pacific Ocean. Their original protocol involved the (1) isolation of bulk genomic DNA from picoplankton collected by tangential flow filtration, (2) fragmentation, size-fractionation (10–20 kb) and cloning of the mixed-population DNA into bacteriophage lambda, (3) screening of the resulting library of recombinant phages by hybridization with 16 S rDNA probes, (4) sequencing of PCR-amplified 16 S rDNA from positive clones, and (5) analysis of the DNA sequence from unique clones to database entries to reveal some of the uncultured diversity of picoplanktonic life in the Pacific Ocean. This series of subsequent steps (DNA isolation, cloning, library screening, sequencing of interesting clones, and DNA comparison) is in essence the classical metagenomic strategy as defined by Handelsman et al. (1998). This basic theme of metagenomics is also depicted schematically in Fig. 1. This figure will be used as a framework to describe the metagenomic methodology and illustrate the many variations that have evolved over the past few years. For good overviews on the subject of metagenomics, several recent reviews are available (Daniel 2005; Deutschbauer et al. 2006; Green and Keller 2006; Handelsman 2004; Kowalchuk et al. 2007; Schloss and Handelsman 2005; Schmeisser et al. 2007; Streit and Schmitz 2004; Tringe and Rubin 2005; Ward 2006; Xu 2006). The present article presents an overview of metagenomics that is to some extent biased towards the discussion in the final section about how to capitalize on metagenomics as a tool in the study of PGPR.

Isolation of metagenomic DNA

The first and obviously most important step in any metagenomic approach is the isolation of DNA from

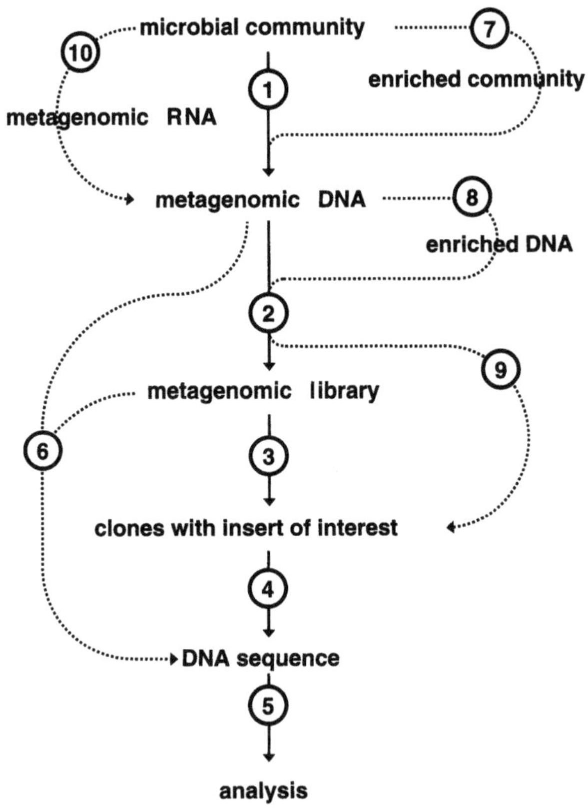

Fig. 1 Schematic representation of the classical metagenomic protocol (*1–5*) and variations on a theme (*6–10*). Each of the following steps is discussed in more detail in the text: (*1*) Isolation of metagenomic DNA; (*2*) Cloning of metagenomic DNA; (*3*) Metagenomic library screening; (*4*) DNA sequencing; (*5*) Sequence analysis; (*6*) Environmental shotgun sequencing; (*7*) Enrichment of a particular subpopulation; (*8*) Enrichment strategies at the DNA level; (*9*) Direct all-or-none selection for clones of interest; (*10*) Isolation of metagenomic RNA

the environment under study. Looking at the many protocols that have been published to date (for a fairly broad overview, see individual chapters in Kowalchuk et al. 2004), it becomes apparent that no single protocol is suitable for the extraction of DNA from all environments. Key issues to consider at this stage are the quantity, purity, integrity, and representativeness of the DNA after isolation. DNA can be extracted from microorganisms by lysis either directly in the environmental sample, or indirectly, i.e. after separation and concentration of the microbial cells from their environmental matrix. The latter is often inevitable or recommended for many environments but this may be for different reasons. Isolation of DNA directly from ocean water is not practical given the low microbial density, so some filtration step is usually performed to first concentrate cells (Fuhrman et al. 1988). Many soil types are notorious for the presence of contaminants such as polyphenolics that co-purify with DNA and can inhibit subsequent steps in the metagenomic process (Tebbe and Vahjen 1993). To prevent this, several groups have developed ways to first separate cells from the soil matrix, e.g. by application to a Nycodenz cushion (Lindahl and Bakken 1995). Others preferentially extract DNA directly from soil, mainly for reasons of increased DNA yield and lesser bias (see below). In these cases, a separate DNA purification step is often included to minimize contamination with unwanted soil substances (Zhou et al. 1996). For direct extraction/purification of environmental DNA from soil, many commercial kits are now also available, for example the SoilMaster™ DNA Extraction Kit (Epicentre, Madison, WI) and PowerMax™ Soil DNA Isolation Kit (Mo Bio Laboratories, Carlsbad, CA).

There is a growing recognition that different DNA extraction methods can yield different results in terms of microbial representation in the final DNA sample (Kozdroj and van Elsas 2000). For example, indirect DNA extraction from marine sediments resulted in a reduced observed microbial diversity at the DNA level than when direct lysis methods were used (Luna et al. 2006). As noted elsewhere (Tringe and Rubin 2005), the harsh lysis methods that are necessary to extract DNA from one organism will cause degradation of the DNA from another organism. A practical implication of this is that more than one type of extraction method should be used in the construction of a metagenomic library. Some DNA extraction protocols have a requirement for the minimal amount of starting material, which can be a disadvantage for the isolation of metagenomic DNA from environments that permit only small sampling sizes. A related issue is the efficiency of DNA extraction. Chemical extraction is usually less effective than procedures that involve rigorous bead-beating (Miller et al. 1999). If DNA yield is a problem – for example, when it is less than the 0.5- to 4-μg minimum to construct a library for shotgun sequencing (Abulencia et al. 2006) – one could consider the use of recently developed protocols for the amplification of whole genomes from environmental samples. One example is the use of Φ29 DNA polymerase which by a process of multiple displacement can provide enough DNA from only five bacterial cells for subsequent analyses (Abulencia et al. 2006).

A final factor to consider is integrity of the isolated DNA which dictates how it can be used in the subsequent steps of the metagenomic protocol. Bead-beating-type procedures tend to fragment DNA and are therefore mostly used for the construction of small-insert libraries, e.g. for shotgun sequencing. In contrast, gentle chemical lysis of cells recovered from soil in agar plugs can produce very large clonable DNA with fragment sizes exceeding 1 Mbp (Berry et al. 2003). The latter is more interesting when the objective is to screen metagenomic libraries for phylogenetic anchors or for particular phenotypes (see "Metagenomic library screening").

Cloning of metagenomic DNA

After extraction and purification of metagenomic DNA, it is usually fragmented to the desired size (either enzymatically or by physical force) and size-separated, if necessary, e.g. on agarose gels (Liles et al. 2004). To carry the metagenomic DNA, a variety of vectors is available, including plasmids, bacteriophages, cosmids, fosmids, and bacterial artificial chromosomes (BACs). A common way to describe these vectors is by the size of DNA fragments that they can typically accommodate, i.e. 0.5–2, 7–10, 35–40, and 80–120 kb, respectively (Xu 2006). However, by this definition some vectors would qualify both as fosmids and as BACs. For example, Epicentre's pCC1FOS and pCC1BAC are virtually identical but are sold as fosmid and BAC, respectively. The difference in nomenclature lies not in the nature of the vector, but in the method that is used to introduce vector-ligated metagenomic DNA into the host strain, in this case *E. coli*. With pCC1FOS, this is mediated through lambda packaging, while pCC1BAC is transformed by electroporation. A limitation of the packaging procedure is that there is an upward size-selection against DNA inserts larger than ~40 kb. On the other hand, DNA inserts smaller than ~35 kb are also rejected, so that a typical fosmid library contains very few false-negatives, i.e. clones that lack an insert or contain very small fragments. With BACs, large DNA fragments can be cloned (Rondon et al. 2000), even as large as 1 Mbp (Berry et al. 2003), but it is also more common for a BAC library to be quite variable in insert size, with a strong selection for small insert sizes and only a relatively small subset of clones carrying very large fragments (Rondon et al. 2000).

Insert size is a critical matter in any metagenomic approach. For environmental shotgun sequencing (see "Environmental shotgun sequencing"), insert sizes of 1.5 to 3 kb are preferred, whereas larger inserts are needed if one wants to maximize on the phylogenetic anchoring approach or if one's objective is to clone large operons e.g. those involved in the production of certain antibiotics (Handelsman et al. 1998). Especially in cases such as the latter, where metagenomic DNA of interest is only detected if it is expressed, the choice of host is crucial to success. *Escherichia coli* is the standard in most metagenomic applications, mainly because of ease of manipulation, but it has limited and biased ability in expressing heterologous genes. Gabor et al. (2004a) provided estimates that only 7% of coding sequences from representatives of the class Actinobacteria would actually be expressed in an independent manner in an *E. coli* background, compared to 73% of the genes from Firmicutes origin. Indeed, the use of alternative hosts, such as *Streptomyces* (Courtois et al. 2003; Wang et al. 2000), *Pseudomonas* (Martinez et al. 2004), or *Rhizobium leguminosarum* (Li et al. 2005) can reveal gene activities that would not have been picked up in an *E. coli* library.

A final consideration in the construction of a metagenomic DNA library is the number of clones that is needed to achieve the desired outcome. It has been estimated that 10^6 BAC clones with 100-kb inserts are required to represent the genomes of all the different prokaryotic species present in 1 g of soil (Handelsman et al. 1998), assuming that all species are equally abundant. If this is not the case, $>10^{11}$ BAC clones, or 10 Tbp of DNA, would be required to achieve adequate genomic representation of rare species (Daniel 2005). Hence, cloning the entire microbial metagenome of an environment as complex as soil is not feasible. To illustrate a more modest goal, suppose one sets out to find bacterial antibiotic resistance genes in a metagenomic library. If one such gene occurs in every 100 bacterial genomes in the metagenome (with an average genome size of 5 Mbp), it would take at least 57,500 40-kb inserts to find a single such gene with a 99% probability. This relative low success rate has not only driven the interest in and application of automated handling systems ('picking robots') to construct and screen metagenomic libraries. It has also led to the exploration and development of variations on the classic metagenomic strategy in order to specifically increase the chances of finding clones of interest (see "Enrichment of a particular subpopulation," "Enrichment strategies at the DNA level," " Direct all-or-none selection for clones of interest," and "Isolation of metagenomic RNA," and the corresponding arrows in Fig. 1).

Metagenomic library screening

Two types of screening strategies can be distinguished. Those that are sequence-based capitalize on pre-existing DNA information to find clones in a metagenomic library that carry inserts with sequence similarity to a gene or locus of interest. This is achieved, for example, by colony or plaque hybridization with probes (Knietsch et al. 2003a; Schmidt et al. 1991) or by PCR using a specific primer set (e.g. Courtois et al. 2003). Most common targets of these searches are the genes for subunits of the ribosome (e.g. Liles et al. 2003; Quaiser et al. 2002), most frequently the small subunit or 16 S rRNA gene. These genes serve as phylogenetic anchors to link the identity of a DNA's owner to part of its biology and evolution by analysis of the DNA sequences flanking the rRNA genes. Special mention is also warranted for the screening method LIL-FISH, for large-insert library fluorescent in situ hybridization (Leveau et al. 2004). In essence, it is an activity screening that utilizes FISH to identify clones in a metagenomic library which heterologously express ribosomal RNA genes. Its compatibility with FACS makes it an attractive method for high-throughput screening of libraries for phylogenetic anchors. A final screening strategy that still needs to be fully explored for its application in metagenomics but worthy of mention here is magnetic capture-hybridization (Jacobsen 1995). This technique involves the use of magnetic beads which are coated with a single-stranded DNA probe complementary to a target gene of interest and which can be used to identify fosmids of interest in a library (Hackl et al., unpublished data).

The second type of library screening is based on the expression of genes of interest through detection of their associated phenotypes in a clone library. The importance of a suitable host strain in this context has already been discussed. Preferably, the gene's phenotype is readily detectable and often involves an enzymatic activity on a plate or in a microtiter plate. Recently, several novel screening methods have been described. Two of these, SIGEX and METREX, are based on the use of green fluorescent protein as a reporter of gene activity. SIGEX stands for substrate-induced gene expression and was developed by Uchiyama et al. (2005). It allows a high-throughput screen for catabolic genes by exploitation of the fact that many such genes, e.g. those that code for the degradation of aromatic compounds are commonly induced by the substrate they are targeted to degrade. Hence, by coupling metagenomic DNA to a promoterless *gfp* gene, library clones carrying genes of interest can be readily identified as green fluorescent cells or colonies in the presence of the inducing substrate. The power of SIGEX lies in the fact that it is compatible with fluorescence-activated cell sorting, which makes library screening a matter of seconds or minutes. METREX (Williamson et al. 2005) is a similar strategy that helps identify functionally active library clones based on an intracellular screen for quorum-sensing inducers.

DNA sequencing

Once clones of interest are identified by a functional or sequenced-based screening, the next step usually involves the elucidation of the inserts' partial or complete DNA sequence. Typically, large-insert fosmid clones are subjected to shotgun or transposon sequencing. To get a typical 10× coverage of a 45-kb fosmid (37-kb insert and 8-kb vector), 600 sequence reads of ~750 bp each are needed.

Most of modern sequencing takes place according to the original Sanger method (Sanger et al. 1977) of chain-termination by dideoxynucleotides and subsequent separation of differently sized DNA fragments by capillary gel electroporesis. This method has proven to be very compatible with the high-throughput demand of many (meta)genomic projects. Recently, novel types of sequencing methods have been developed that are faster and have an even greater capacity. For example, pyrosequencing now allows the sequencing of 25 million bases in one 4-h run with an accuracy of 99.96% (Margulies et al. 2005). While this technology will not likely replace traditional sequencing, it has great utility for the environmental shotgun sequencing strategies discussed later (see "Environmental shotgun sequencing"). While there are some serious limitations to pyrosequencing – the main ones being the generation of relatively short (~100-bp) reads and the poor ability of most current DNA analysis programmes to assemble such short reads – it is an exciting new development that will undoubtedly revolutionize the field of microbial (meta)genomics.

Sequence analysis

There are many ways to analyze metagenomic DNA sequences (Chen and Pachter 2005). Gene-finding is supported by software applications such as GeneDB (Meyer et al. 2003), Artemis (Rutherford et al. 2000), Glimmer (Delcher et al. 1999), and FGenesB pipeline (www.softberry.com). These programmes use special algorithms to identify coding sequences, as well as other features such as promoters, terminators, operons, tRNA and rRNA. They can also provide a functional prediction of each identified putative gene based on sequence similarity of its predicted product. The programme MetaGene (Noguchi et al. 2006) is a prokaryotic gene-finding application that was designed specifically for the analysis of metagenomic datasets with many unassembled reads. Such datasets are typical for many types of environmental shotgun sequence projects (see "Environmental shotgun sequencing"). A crude approach to gene-finding is a BlastX-type of analysis, by which the DNA query sequence is first translated into protein sequences in all six reading frames, after which these products are compared against the existing protein databases. This approach depends heavily on the availability of similar sequences in the database and novel genes might be missed in this way.

A particular problem with small-sized metagenomic DNA fragments is that they often lack phylogenetic markers in the shape of e.g. a 16 S rRNA gene. In those cases, parameters such as G+C content, BlastX scores, and codon usage frequencies can also be used as indicators of phylogenetic origin (Chen and Pachter 2005). Analysis of oligonucleotide frequencies is another promising approach to this problem, since these frequencies tend to exhibit species-specific patterns (Abe et al. 2005). TETRA (Teeling et al. 2004) is available as a web-service (www.megx.net/tetra) or stand-alone application that automates the task of comparing tetranucleotide frequencies. It computes correlation coefficients between patterns of tetranucleotide usage in DNA and works best with sequences of about 40 kb. Most recently, TETRA was used to group metagenomic DNA sequences from the marine worm *Olavius algarvensis* into four clusters representing four prokaryotic symbionts (Woyke et al. 2006).

Environmental shotgun sequencing

As already alluded to in the introduction, a recent development in metagenomics is environmental shotgun sequencing, also known as whole-genome sequencing (Chen and Pachter 2005). It usually involves the construction and end-sequencing of small-insert libraries from DNA directly isolated from the environment under study, although several variations exist. For example, the dataset of DeLong et al. (2006) consisted of end-sequences from a large-insert fosmid library, and Edwards et al. (2006) applied pyrosequencing which makes library construction redundant altogether. Examples of environments from which prokaryotic communities have been shotgun-sequenced are the Sargasso sea (Venter et al. 2004), the North Pacific ocean's surface to near-sea floor depths (DeLong et al. 2006), sunken whale carcasses (Tringe et al. 2005), deep-sea sediment (Hallam et al. 2004), an acid mine biofilm (Tyson et al. 2004), groundwater in banded iron formations of a subsurface mine (Edwards et al. 2006), the human distal gut (Gill et al. 2006), drinking-water networks (Schmeisser et al. 2003), and an agricultural soil (Tringe et al. 2005).

One characteristic that all environmental shotgun sequencing projects share is the incredible amount of data that is generated (Schloss and Handelsman 2005). The Sargasso Sea project alone produced 2 million sequence reads, or 1,600 Mbp of DNA sequence (Venter et al. 2004). This explosion in environmental sequence data led several authors (Handelsman 2005; DeLong 2004) to refer to metagenomics as 'mega-genomics' or 'mega-metagenomics,' respectively. This 'genomics on a massive scale' poses considerable challenges for data assembly and annotation. Successful assembly of individual reads into contigs is inversely proportional to the complexity of the prokaryotic community from which the DNA originated. The acid mine biofilm studied by Tyson et al. (2004) represented a relatively simple community, with only three bacterial and three archeal lineages. From two members of the biofilm, *Ferroplasma* type II and *Leptospirillum* group II, near-complete genome sequences could be obtained after assembly. In more complex environments, this is typically not the case. For example, 50% of the reads from the Sargasso Sea could not be assembled, and for the agricultural soil, with an estimated species richness of >3,000 (Tringe et al. 2005), the percentage of unassembled reads even approached 100%. Interestingly, Tringe et al. (2005) recently questioned the need for assembly in such cases altogether and introduced the concept of environmental gene tags (EGTs), which takes a more gene-centric approach to the analysis of environmental sequencing. EGTs are in essence annotated individual reads from a metagenome shotgun project. Predicted genes on these EGTs are derived from individual members of the microbial community under study, and could potentially code for a habitat-specific adaptation. By comparison of EGT 'fingerprints' of different environments, Tringe et al. (2005) indeed observed emerging patterns of habitat-specific gene functions, for example, an over abundance of rhodopsin-like proteins in ocean surface waters, enzymes for the degradation of plant material in soil, and sodium transport proteins in marine environments. Basically, this approach represents an in silico version of substractive hybridization (Schloss and Handelsman 2005) and presents one of several examples (Rodriguez-Brito et al. 2006) on how to tackle the challenges of comparative analysis of metagenomic datasets.

Enrichment of a particular subpopulation

This section is the first of four that describe some of the variations that have evolved from the basic metagenomic theme (Fig. 1) with the specific goal to increase the chance of finding a gene or function of interest. One of those variations is the subject of this section and involves community fractionation, i.e. the isolation of DNA only from a selected subpopulation of the environmental sample under study. One such type of selection is size. Using filters with different pore sizes, bacteriophage communities can be selectively enriched for metagenomic analyses (Edwards and Rohwer 2005). Similarly, size fractionation was used to select for consortia > 3 μm from deep-sea sediments (Hallam et al. 2004). Other strategies for enrichment of subpopulations include affinity purification or differential lysis (Tringe and Rubin 2005).

A very effective type of selection at the community level is based on the application of selective pressure to the environmental sample under study in order to favour the growth of bacteria expressing a desired activity. In an early example of this strategy, Healy et al. (1995) readily recovered clones expressing cellulase and xylosidase activities from libraries of DNA isolated from a mix of thermophilic, anaerobic digesters, that were in continuous operation with lignocellulosic feedstocks for >10 years. In another case, Gabor et al. (2004b), DNA was isolated from enrichment cultures in which amides (either singly or as a mixture of aromatic and non-aromatic forms) were supplied as the sole nitrogen source to obtain a library enriched for amidases with different substrate specificities. Similarly, Entcheva et al. (2001) enriched for biotin-producing bacteria to isolate new biotin biosynthesis operons, whereas Knietsch et al. (2003b) pre-selected for utilization of glycerol and 1,2-propanediol to metagenomically identify genes encoding alcohol oxidoreductases.

Another clever selection procedure allows the isolation of DNA only from the live fraction of cells in a community. Nocker and Camper (2006) reported that propidium monoazide (PMA), like propidium iodide, is highly selective in penetrating only 'dead' bacterial cells where it intercalates in the DNA and can be covalently cross-linked by exposure to bright light. This process renders the DNA insoluble and results in its loss during subsequent genomic DNA extraction. Subjecting a bacterial population comprised of both live and dead cells to PMA treatment would thus result in selective removal of DNA from dead cells, which can be useful if one is interested only in analyzing the currently active, living fraction of a microbial community.

Enrichment strategies at the DNA level

A second kind of selection for genes or gene functions of interest involves selection at the level of metagenomic DNA after it has been isolated from the environment under study. The simplest application of this type of enrichment is the use of metagenomic DNA as a template in PCR (with degenerate primers, if needed) to amplify and clone genes of interest (Marchesi and Weightman 2003). A method described by Nesbø et al. (2005) allows for the enrichment only of metagenomic DNA that carries phylogenetic markers. Central in this approach is the restriction enzyme I-*Ceu*I, which targets a 19-bp sequence that is conserved in 23 S rRNA of most bacteria. After isolation, fragmentation and end-repair, metagenomic DNA is digested with I-*Ceu*I and ligated to vector pCC1.FOS.CeuI.23S, a derivative of pCC1FOS containing unique I-*Ceu*I and blunt sites. To illustrate the effectiveness of this approach, fosmid libraries were constructed from anaerobic sediments sampled from Baltimore harbour, and found to be enriched by an approximate factor of 80 for clones carrying 23 S rRNA genes. An attractive side advantage of the method is that end-sequencing of clones in this library, which is relatively cheap, can provide instant information on the identity of the DNA's origin based on 23 S similarity.

Stable isotope probing is a technique by which to isolate from the metagenome pool specifically that DNA which is derived from organisms that can metabolize a particular substrate (Friedrich 2006). This is achieved by incorporation of ^{13}C- or ^{15}N-labelled substrates into biomass, including DNA, of the active subpopulation in the community under study. The labelled DNA (from the active microorganisms) is then separated from unlabelled DNA through density gradient centrifugation and used as starting material for the cloning into vectors. One example is the report by Dumont et al. (2006) which describes the application of ^{13}CH4 to forest soil to generate a BAC library enriched for methane monooxygenase genes. Schwarz et al. (2006) used ^{13}C-labelled glycerol on a sediment sample of the

Wadden Sea to enrich by 2.1- to 3.8-fold DNA fragments that carried genes encoding coenzyme B12-dependent glycerol dehydratases.

Galbraith et al. (2004) have described a novel method based on suppressive subtractive hybridization to specifically isolate from one metagenomic DNA sample (designated the 'tester') fragments that are absent in another metagenomic DNA sample (the 'driver'). Applied to the study of microbial populations in the rumen, this approach revealed an unexpectedly large difference in archaeal community structure between steers fed identical diets. It should be noted that there are several drawbacks to this method: e.g. large amounts of subtractive sequences must be read to achieve an accurate sense of the degree of genetic diversity between two samples. However, the technique is very suitable for comparative purposes, able to zoom in on relatively small differences, and allows to possibly link those differences in genetic diversity to physical, chemical or biological dissimilarities between environmental samples.

Direct all-or-none selection for clones of interest

A very powerful instrument in the metagenomic toolbox is the use of conditional survival of the metagenomic host through functional complementation. In essence, one exploits the fact that the host that is being used needs a particular type of gene or gene cluster from the pool of metagenomic DNA fragments in order to survive an imposed condition. This concept has been applied by several groups. For example, Li et al. (2005) screened a metagenomic library for clones that could correct tryptophan auxotrophy in hosts *E. coli* and *R. leguminosarum*, and in doing so identified several different *trp* operons. Gabor et al. (2004b) were successful in finding metagenomic clones that complemented the leucine auxotrophy of the host strain *E. coli* TOP10 host by expression of amidase activity on a medium containing phenylacetyl-L-leucine or D-phenylglycine-L-leucine as the sole source of leucine. Similarly, Entcheva et al. (2001) used a biotin-auxotroph *E. coli* strain to pick up several clones with biotin biosynthesis operons.

The major advantage of this approach of complementation is its all-or-none character: only if a gene is present and expressed will the host survive, so that false-positives can be expected to be rare. False-negatives, on the other hand, may be more frequent, depending on the host or range of hosts that is being used. Li et al. (2005) identified at least one set of *trp* genes that complemented *R. leguminosarum* but not *E. coli*, while Entcheva et al. (2001) noted that all of the biotin operons that were recovered with *E. coli* as a host had highest similarity to similar operons in *Enterobacteriaceae*. Thus, host choice greatly determines the success rate of this strategy in finding genes and gene functions of interest.

Isolation of metagenomic RNA

A final variation on the classical metagenomic theme starts with the isolation of RNA, not DNA, from the environment under study, followed by reverse transcription of this RNA, and cloning of the resulting cDNA. This approach can offer answers not to the question 'who is out there?' but to 'who is active out there?'. It is a technically challenging approach, mostly because of RNA instability. Working protocols are available for the isolation of environmental prokaryotic RNA (Hurt et al. 2001) and cloning of short cDNA sequences (Poretsky et al. 2005). Grant et al. (2006) recently described procedures for stabilizing eukaryotic RNA in environmental samples in the field such that they can be transported back to the laboratory, the RNA isolated, and cDNA libraries made for subsequent sequencing and expression studies. Mills et al. (2004) were able to determine the composition of the metabolically-active fraction of microbial communities in marine sediments by rRNA extraction and reverse-transcription to obtain clonable complementary 16 S ribosomal DNA.

Metagenomic stories of success and words of caution

The present excitement about metagenomics has its roots in a number of clear success stories (Table 1). One is the discovery of a member of a new class of rhodopsins, encoded on a BAC-insert from an uncultivated marine α-Proteobacterium (Béjà et al. 2000b). Proteorhodopsin is a retinal-binding bacterial integral membrane protein that acts as a light-driven proton pump in surrogate host *E. coli* (Béjà et al. 2001). Subsequent studies have revealed the abun-

Table 1 A selection of success stories and milestones in the field of metagenomics

Origin of metagenomic DNA	Total DNA cloned/ sequenced	Average insert size (kb)	Type of screening (sequence-based, activity, shotgun sequencing)	Number of clones screened/ sequenced	Number of positive clones (%)/sequence assembly information	Success and/ or milestone	Reference
Concentrated picoplankton (2×10^{12} cells) from 8 m^3 of Pacific Ocean water	1–2×10^8 kb in 10^7 clones	10–20	Sequenced-based (Southern hybridization and PCR) for ribosomal RNA genes	3.2×10^4	38 (0.1%)	First real application of the metagenomics principles	Schmidt et al. 1991
Two thermophilic, anaerobic digesters	>24 Mbp	1.6–10.5	Screening for a) carboxymethylcellulase, b) 4-methylumbelliferyl β-d-cellobiopyranoside activity	15,000	(a) 12 (0.08%), (b) 11 (0.07%)	First example of screening metagenomic libraries for activity	Healy et al. 1995
soil from a meadow, sugar beet field, and valley of the river Nieme (Germany)	>4.65 Mbp	5–8	Screening of *E .coli* library clones on tetrazolium indicator plates containing 4-hydroxybutyrate	930,000	5 (0.0005%)	First report on the isolation of a biocatalyst from soil-derived libraries	Henne et al. 1999
Soil from the West Madison Agricultural Research Station in Madison, WI	100 Mbp in library SL1, 1 Gbp in library SL2	27–45	(a) PCR screening for rRNA genes, (b) DNase, lipase, amylase and antibacterial activity, (c) hemolytic activity, (d) cellulase, chitinase, esterase, keratinase, protease, hemolytic activity/ siderophore production	(a) 3,648 (SL1), (b) 3,648 (SL1), (c) 24,576 (SL2), (d) 3,648 (SL1)	(a) 7 (0.2%), (b) 12 total (0.3%); antibacterial clone SL1-36C7 was completely sequenced, (c) 29 (0.1%), (d) 0 (<0.03%)	Key paper by the group that gave the field its name (Handelsman et al. 1998), demonstrating the enormous potential of large-insert metagenomic libraries for finding new activities	Rondon et al. 2000
Unspecified	Unspecified	Unspecified	Screening of library clones with HPLC-ESIMS to identify transformants producing novel compounds	1,020	2 (0.2%); clone 436-s4-5b1 produced 4 types of terragin and 446-s3-102g1 another type	One of the first reports of the use of a non-*E. coli* host (i.e. *Streptomyces lividans*) for screening of metagenomic libraries	Wang et al. 2000

Table 1 (continued)

Origin of metagenomic DNA	Total DNA cloned/ sequenced	Average insert size (kb)	Type of screening (sequence-based, activity, shotgun sequencing)	Number of clones screened/ sequenced	Number of positive clones (%)/sequence assembly information	Success and/ or milestone	Reference
Natural microbial assemblage collected from surface waters off the Californian coast	468 Mbp	60–80	PCR screening for the presence of (a) archaeal 23 S ribosomal RNA, (b) SAR11-like rRNA, and (c) bacterial 16S-ITS-23S sequences	6,240	(a) 1 (0.02%), (b) 3 (0.05%), (c) ≥22 (≥0.4%); several clones were fully sequenced and on one (BAC clone 31A08) the proteorhodopsin gene was found	Discovery of a bacterial rhodopsin gene and a previously unrecognized mode of light-driven energy generation in bacteria	Béjà et al. 2000a, 2000b
Soil (source or type not specified)	Unspecified	Unspecified	Screening for growth inhibition of *Bacillus subtilis,* indicating the production of antibacterial activity	700,000	6 5 (0.009%); one clone, cosmid CSL12, was fully sequenced	First instance of isolating the genetic information for antibiotic production from a metagenomic library	Brady and Clardy 2000
Soil from a meadow, sugar beet field, and valley of the river Nieme (Germany)	>7.4 Gbp	5–8	Screening for growth rescue of an antiporter-deficient *E. coli* host strain by metagenomic DNA fragments	1,480,000	2 (0.0001%)	First example of using functional complementation in metagenomics to discover new genes (coding for antiporter activity in this case)	Majernik et al. 2001
200 l of surface seawater from S cripps Pier (SP) and the Mission Bay (MB)	Unspecified	0.7	Shotgun sequencing	1,934 sequences	3.5% of sequences could be assembled	One of the first shotgun sequencing approaches, showing tremendous, previously uncharacterized, viral diversity	Breitbart et al. 2002

Upper layer (0–5 cm) of a calcerous grassland near Darmstadt (Germany)	0.9 Gbp	32.5–43.5	PCR on pooled clones using archaea-specific 16S rDNA primers followed by colony hybridization of clones within positive pool with labeled PCR product	25,278	1 (0.004%): clone 29i4 was fully sequenced	First full sequence of an archaeal metagenomic clone	Quaiser et al. 2002
Microbial biomass on the surface of the rubber-coated valves in the drinking water networks in the state of North Rhine-Westphalia (Germany)	Unspecified	1–5	Shotgun sequencing	5,000 sequences	2,496 high-quality reads representing a total of >2 Mbp sequence	First report of shotgun sequencing of a microbial community	Schmeisser et al. 2003
Biofilm growing on the surface of flowing acid mine at Iron Mountain, CA	76.2 Mbp	3.2 kb	Shotgun sequencing	103,462 reads with average length of 0.7 kb	Assembly into 10.83 Mbp	Near-complete genomes of dominant members of a simple microbial community	Tyson et al. 2004
Surface water samples from Sargasso sea	1.63 Gbp	2–6	Shotgun sequencing	2 million reads with average length of 0.8-kb	Assembly into 1 Gbp of nonredundant sequence	'Megagenomics': unprecedented sequence-based discovery of 148 new bacterial phylotypes and 1.2 million novel genes	Venter et al. 2004
Soil from the West Madison Agricultural Research Station in Madison, WI	5.4 Gbp	2.7–45	Screening for acquired resistance in host strain *E.coli* to (a) kanamycin, (b) tetracycline, apramycin, butirosin, tobramycin, and (c) nalidixic acid	1,186,200	(a) 5 (0.0004%), (b) 1 each (0.00008%), (c) 0 (<0.00008%)	Shows the use of positive selection to isolate rare gene activities in a metagenomic library	Riesenfeld et al. 2004a

Table 1 (continued)

Origin of metagenomic DNA	Total DNA cloned/sequenced	Average insert size (kb)	Type of screening (sequence-based, activity, shotgun sequencing)	Number of clones screened/sequenced	Number of positive clones (%)/sequence assembly information	Success and/or milestone	Reference
Surface soil (0–10 cm) from a farm in Waseca County, MN	100 Mbp	2.4 kb	Shotgun sequencing	149,085 reads	100% of reads unassembled	Metagenomic sequencing of complex microbial community (soil); first demonstration of quantitative comparative metagenomics	Tringe et al. 2005
Unspecified forest soil	>23 Mbp	10–30	Screening by hybridization to *pmoA* gene (particulate methane monooxygenase)	2,300	1 (0.04%)	Proof-of-concept for the use of stable isotope probing to generate metagenomic library from specific subpopulation	Dumont et al. 2006
Sub-surface banded iron formation from Soudan Iron Mine in Minnesota	70 Mbp	n/a	Shotgun pyrosequencing	723,013 reads with average length of 0.1 kb	No assembly	First application of pyrosequencing to environmental samples	Edwards et al. 2006
Fecal material from two healthy human subjects	Unspecified	2–3	Shotgun sequencing	139,521 reads with average length of 0.9 kb	33.8 Gbp assembled sequence plus 45 Gbp singletons	Metagenomic analysis of the human distal gut microbiome	Gill et al. 2006

dance and diversity of this type of protein in ocean waters (Venter et al. 2004). This discovery clearly illustrates the utility of metagenomics to reveal unsuspected biological functions. Another story of success is the Sargasso Sea project (Venter et al. 2004). It demonstrated, unambiguously, the muscle of metagenomics: by sheer brute force, 1 Gbp of nonredundant sequence was generated, two to three orders of magnitude more than the human genome (Venter et al. 2001). From <2 cubic meters of ocean water, an unprecedented microbial diversity was uncovered, including 148 previously unknown bacterial phylotypes and 1.2 million previously unknown genes. The Sargasso Sea study greatly stimulated the discussion on the limits and the future of metagenomics and still serves as a reference point for many of the environmental shotgun sequencing projects that followed in its footsteps. A third tale of metagenomic fame is that of the acid mine drainage biofilm (Tyson et al. 2004). It actually makes several interesting points, most of which are a direct consequence of the low-level complexity of the community under study. Due to the latter, the genomes of two members in the consortium could be reconstructed to near completion, enabling a true systems-biology approach to studying the interplay of microbial metabolism and mineral dissolution (Allen and Banfield 2005). A proteomic analysis of this community showed that biofilm polymer production and nitrogen fixation appeared to be partitioned among community members and that an abundant cytochrome-like protein might be essential to the production of acid mine drainage (Ram et al. 2005). The metagenome sequence also revealed single-nucleotide polymorphisms which prompted addressing the biofilm community in terms of population genomics and evolutionary ecology (Whitaker and Banfield 2006). Furthermore, based on clues from its metagenomic sequence, one of the previously unculturable consortium members, *Leptospirillum ferrodiazotrophum*, could be grown in the laboratory (Tyson et al. 2005), reducing the number of 'unculturable' bacteria by one.

Many of the stories of metagenomic success also contain important words of caution. In hindsight and in words of DeLong (Sreenivasan 2001), the discovery of the proteorhodopsin gene was to a large degree a matter of serendipity, or luck. Since it was one of the very first successful cases to demonstrate the ability of metagenomics to link phylogeny and function, the bar was put high for any future endeavour in this direction. In practice, while the chance of finding novel genes on any metagenomic fosmid- or BAC-insert is not very low, it is much less probable to find a gene coding for a never-seen-before prokaryotic life-style such as is the capture of light for energy. Unrealistic expectations may also arise from a false sense of probability of finding a particular gene of interest in a metagenome library. Even without considering potential problems of expressing heterologous DNA in a surrogate host or insufficient homology to identify clones using PCR or hybridization, it remains a laborious and often underestimated task to screen through many thousands of clones, especially in the case of metagenomic DNA from environments with high microbial complexity. For example, it took 1,186,200 clones containing a collective 5.4 Gbp of soil DNA to identify nine unique clones that conferred resistance to aminoglycoside antibiotics and one clone expressing resistance to tetracycline (Riesenfeld et al. 2004a). A hit rate of one interesting gene per 5–5,000 Mbp of cloned metagenomic DNA is not uncommon for activity-based screenings (Lorenz and Eck 2005) and should probably be considered a normal operating range in any metagenomic screening.

A final consideration in this section is the recent argument (Oremland et al. 2005) that the predictive and interpretative value of metagenomics is limited by the validity of database entries. Many genes are listed with unknown function and for many genes that have a function assigned based on sequence similarity, this function still needs to be validated experimentally. In this context, it is of extreme value to have culturable representatives available from the environment under study. The Sargasso Sea study and other massive-scale shotgun projects clearly show the rapid progress in sequencing capacity, but do not contribute greatly to our ability to assign gene function. Metagenomics is more than a descriptive science; it should be appreciated as a methodology that is complementary to conventional approaches in testing hypotheses on the composition, diversity and functionality of microbial communities.

Metagenomics and the study of plant growth-promoting rhizobacteria

It was mentioned earlier in the introduction that the first and most important step in a metagenomic approach is the isolation of DNA from the environment under study. This is not quite right: it is probably more important to first ask whether it is warranted or wise to invest in a metagenomics approach at all and for what purpose. The remainder of this article will be an attempt to address this question in general terms for the study of plant growth-promoting rhizobacteria (PGPR).

The term PGPR (Kloepper and Schroth 1978) refers to those plant root ('rhizosphere')-associated bacteria that are capable of stimulating plant growth, e.g. by improving plant nutrition, by the production of plant growth regulators or by preventing the attack of pathogenic microorganisms. PGPR vary in their degree of intimacy with the plant, from intracellular, i.e. existing inside root cells, to extracellular, i.e. free-living in the rhizosphere (Gray and Smith 2005). Some PGPR are commercially available as inoculants and have applicability for example in agriculture, forest regeneration, and phytoremediation of soils (Lucy et al. 2004).

There is a clear potential for metagenomics to contribute to the study of microbial communities of the rhizosphere, in particular PGPR. Possible contributions include (1) the discovery of novel plant-growth promoting genes and gene products, and (2) the characterization of (not-yet-)culturable PGPRs. Before discussing these in more detail below, it is worth noting that in practical terms, the application of metagenomics to PGPR in the rhizosphere greatly benefits from previous advances in DNA isolation and library construction from other environments. Rhizosphere soil poses more or less the same challenges as bulk soil, for which several metagenomic success stories have been published (Daniel 2005). Probably the biggest obstacle in the construction of a metagenomic library from rhizosphere soil DNA is the relative low availability of starting material. To 1 cm of root typically adheres only 20 mg of soil (Jacobsen 2004), so one needs (depending on the plant species under investigation) 50 to 500 cm of root material in order to apply a DNA extraction method that requires 1 to 10 g of soil. Several protocols have also been developed for the isolation of metagenomic bacterial DNA from inside plant material. For example, Jiao et al. (2006) describe an indirect method based on enzymatic hydrolysis of plant tissues to release associated microorganisms for subsequent DNA isolation and cloning. While optimized for leaves and seeds, this method seems readily adaptable for use with root material, and thus of great use to the metagenomic exploration of microorganisms in the rhizosphere.

Novel plant growth-promoting genes and gene products

For many of the traits or mechanisms known to be plant growth-promoting (Bloemberg and Lugtenberg 2001), in vitro activity assays have been described and are, at least in theory, exploitable for gain-of-function screenings of a metagenomic library from rhizosphere DNA. For example, antibiotic activity towards (plant-pathogenic) bacteria or fungi can be assessed by testing whole-cell library clones or their extracts in a medium- or high-throughput manner for performance in confrontation assays. Many of such assays have been described using a variety of indicator strains, including some of the most important soil borne pathogens, e.g. the bacteria *Erwinia* (Emmert et al. 2004) and *Xanthomonas* (Rangarajan et al. 2003), the fungi *Fusarium* (Chin-A-Woeng et al. 1998; Kim et al. 2006), and *Rhizoctonia* (Kim et al. 2006; Rangarajan et al. 2003), and the fungus-like oomycetes *Phytophthora* (Kim et al. 2006) and *Pythium* (Kim et al. 2006; Rajendran et al. 1998). Production of the plant hormone indole 3-acetic acid (IAA) by metagenomic library clones can be measured using high-pressure liquid chromatography or colorimetric assays (Bric et al. 1991; Omer et al. 2004; Radwan et al. 2002; Leveau and Lindow 2005), while cytokinins and their metabolites are detectable in supernatants by e.g. immunoaffinity chromatography (Timmusk et al. 1999). Genes for nitrogen fixation are retrievable with the use of nitrogen-free media (Ding et al. 2005; Hashidoko et al. 2002; Tejera et al. 2005). Similarly, genes for the utilization of particular rhizosphere exudates could be recovered using an all-or-none complementation selection for growth on minimal medium containing these exudates as sole source of energy, carbon and/or nitrogen. In a similar approach, clones expressing 1-aminocyclopropane 1-carboxylate (ACC) deaminase, a plant growth-promoting enzyme that lowers plant ethylene levels (Glick et al. 1998), could be selected for by using ACC as sole source of nitrogen, as described previously (Holguin and Glick 2003; Shaharoona et al. 2006). The activities of lytic enzymes are most easily identified through clear zones around colonies on solid media, as has been documented e.g. for several biocontrol chitinases (Basha and Ulaganathan 2002; Gohel et al. 2004; Kobayashi et al. 2002; Leveau et al. 2006). Assays based on halo formation are also available to identify PGPR-related phenotypes such as solubilization of mineral phosphate (Rodriguez et al. 2000) and siderophore production (Lee et al. 2003).

For the functional screening of library clones for PGPR functions, the use of alternative hosts seems very promising and rational, for several reasons. First, there is an abundant availability of phylogenetically diverse culturable PGPRs (Vessey 2003, Lucy et al. 2004) which could improve the probability of finding genes of interest, especially those that are not expressed in *E. coli* and whose full activity requires a specific PGPR background. The host role could also be played by several of the numerous defined mutants of PGPR that carry a knockout in one or several genes contributing to a particular PGPR phenotype. Such mutants could be useful to screen libraries for heterologous genes and gene functions by a functional complementation approach. Proof-of-principles for such an approach are available in the literature, e.g. single-gene complementation of a mutant of *Burkholderia* sp. strain PsJN in quinolinate phosphoribosyltransferase (QAPRTase) activity (Wang et al. 2006), a mutant of *Pseudomonas putida* WCS358 unable to produce the antibiotic pseudobactin 358 (Devescovi et al. 2001), and a mutant of *Pseudomonas chlororaphis* PCL1391 impaired in the production of the antifungal secondary metabolite phenazine-1-carboxamide (PCN; Girard et al. 2006). Especially interesting in this respect is the use of mutants that lack one or more genes in a multi-gene pathway for the production of antibiotics by enzymes such as polyketide synthases (Staunton and Weissman 2001) and non-ribosomal peptide synthases (Raaijmakers et al. 2006). The modularity underlying such proteins allows for a strategy of combinatorial complementation (Coeffet-Le Gal et al. 2006), possibly leading to the discovery of antimicrobials with new structures and new target specificities (Wenzel and Müller 2005).

Activity screenings such as the ones described above have the potential to retrieve never-before-seen genes with PGPR activity from the metagenomic pool. In contrast, the metagenomic harvest from sequence-based approaches such as PCR and/or Southern hybridizations will inevitably uncover only genes that match the specificity of the primers and/or probes that were used to find them. Nevertheless, screening rhizosphere DNA for PGPR-related genes by PCR (e.g. Juraeva et al. 2006; Sato et al. 1997) or Southern hybridization (e.g. Blaha et al. 2005; Shah et al. 1998) has several advantages in a metagenomic setting. Most importantly, sequencing the flanking regions of such genes on large-insert fosmid or BAC clones could provide insight into the identity of their owner, on the genetic context of these PGPR genes and possibly on the mechanisms of their regulation.

Highly complementary to activity- and sequence-based screenings, a third approach to finding novel plant growth-promoting genes and gene functions is through comparative metagenomics. For one, the rhizosphere can be viewed as an environment that in comparison to the bulk soil is enriched in particular types of microorganisms, including PGPR. Indeed, there is ample evidence that the microbial diversity as measured by phylogenetic markers such as ribosomal RNA genes can differ dramatically between bulk and rhizosphere soil (Costa et al. 2006a; Sanguin et al. 2006). EGT fingerprinting by shotgun sequencing (Tringe and Rubin 2005) or suppressive subtractive hybridization (Galbraith et al. 2004) of bulk and rhizosphere soil compartments could reveal differences in the type of gene adaptations that each compartment selects for. It is expected that genes with PGPR-like functions would be enriched in the rhizosphere library. Similarly, comparison of the genomic diversity of disease-suppressive and non-suppressive soils (Weller et al. 2002) could expose genetic factors that contribute to or are predictive of the suppressiveness towards e.g. pathogenic micro-organisms or nematodes.

Characterization of (not-yet-)culturable PGPRs

Due to a historical bias to study those microorganisms that can be grown in the laboratory, there is limited knowledge on the abundance and activity of not-yet-culturable PGPR. However, there are several examples of their existence and contribution to plant health, e.g. *Pasteuria penetrans*, a not-yet-culturable bacterium parasitic to plant-pathogenic nematodes (Fould et al. 2001), the nitrogen fixing activity by viable-but-not-culturable *Azoarcus* grass endophytes (Hurek et al. 2002), and the obligate biotrophism of arbuscular mycorrhizal (AM) fungi (Millner and

Wright 2002). Bacteria belonging to the *Acidobacteria* and *Verrucomicrobia* are in many rhizospheres among the most abundant, difficult-to-culture representatives (Buckley and Schmidt 2003; Gremion et al. 2003). However, it is not clear if and how their abundance is correlated to their contribution towards plant health. A phylogenetic anchoring approach, e.g. using the previously described I-*Ceu*I method (Nesbø et al. 2005) to construct a library over-represented in DNA fragment harbouring 23 S rRNA genes, in combination with a PCR-based screening of this library with Acidobacterial or Verrucomicrobial primers would allow a (partial) insight into the genomes of these bacteria beyond the limited dataset that currently exists for these classes of bacteria (Liles et al. 2003; Quaiser et al. 2003; Wagner and Horn 2006) and into their possibly beneficial effect on plant growth. Major progress is being made in the development of new cultivation techniques (Joseph et al. 2003; Kaeberlein et al. 2002; Stevenson et al. 2004; Zengler et al. 2002), which offers the prospect that many more of the formulated hypotheses based on metagenomic analysis of the rhizosphere and its PGPR constituency (see below) will become testable as the number of culturable representatives from the rhizosphere steadily increases.

An analysis of the rhizosphere by comparative metagenomics holds the promise to reveal several important questions regarding the unculturable fraction of the rhizosphere community. For one, it could expose what actually constitutes this fraction from a comparison of metagenomic DNA isolated directly from rhizosphere to DNA isolated from all the colonies forming on solid media after plating from that same rhizosphere (i.e. the culturable fraction). One could expect a phylogenetic analysis of these two libraries to show differences, based on previous observations (e.g. Sliwinski and Goodman 2004; Costa et al. 2006b), and with large-scale DNA sequencing of both libraries, a start could be made to contrast the genetic diversity of the two populations. Furthermore, by comparison of the functions enriched for in a library from rhizosphere soil versus one from bulk soil, the degree of the selection in each of the compartments for particular microbial activities, specifically those with PGPR relevance, can be estimated. A shotgun sequencing approach for unlocking the unculturable diversity of rhizosphere bacteria, including PGPR, has not yet been reported. Recently, a study was published (Erkel et al. 2006), describing the use of metagenomic sequencing to reconstruct the 3.18-Mbp genome of rice cluster I (RC-I) Archaea with origin in the rice rhizosphere. DNA for the shotgun library was isolated from a methanogenic enrichment culture using rice paddy soil as an inoculum. While RC-I Archaea do not necessarily qualify as PGPR, the study shows that shotgun sequencing in combination with a prior enrichment strategy towards an originally complex rhizosphere population allows the metagenomic analysis of rhizobacteria with a particular function of interest.

Conclusions

In summary, the tools of metagenomics offer many openings into a broadened view of the rhizosphere in general and of PGPR and their activities in particular. Several existing assays for PGPR activity have been listed here and proposed to have immediate utility for the screening of large-insert DNA libraries for gain-of-function phenotypes. The discovery of novel PGPR activities, either by functional screening or based on DNA sequence information, will add enormously to our understanding of the mechanistic variation that exists in PGPR phenotypes. It will also benefit our ability to improve existing PGPR, by adding to the pool of exploitable PGPR genes and utilization of this pool to develop PGPRs with enhanced performance (Downing and Thomson 2000; Glick and Bashan 1997; Holguin and Glick 2003; Timms-Wilson et al. 2004). The use of metagenomics in parallel with established or novel molecular approaches to the study of PGPR, such as genome sequencing of new PGPR isolates (Jeong et al. 2006) and transcriptional profiling of PGPR (Mark et al. 2005; Wang et al. 2005) will undoubtedly lead to the discovery of novel mechanisms of PGPR activity, new types of PGPR identity and a fresh look on the biology and practical application of PGPR.

References

Abe, T., Sugawara, H., Kinouchi, M., Kanaya, S., & Ikemura, T. (2005). Novel phylogenetic studies of genomic sequence fragments derived from uncultured microbe mixtures in environmental and clinical samples. *DNA Research*, *12*, 281–290.

Abulencia, C. B., Wyborski, D. L., Garcia, J. A., Podar, M., Chen, W. Q., & Chang, S. H., et al. (2006). Environmental whole-genome amplification to access microbial populations in contaminated sediments. *Applied and Environmental Microbiology, 72*, 3291–3301.

Allen, E. E., & Banfield, J. F. (2005). Community genomics in microbial ecology and evolution. *Nature Reviews Microbiology, 3*, 489–498.

Amann, R. (2000). Who is out there? Microbial aspects of biodiversity. *Systematic and Applied Microbiology, 23*, 1–8.

Basha, S., & Ulaganathan, K. (2002). Antagonism of *Bacillus* species (strain BC121) towards *Curvularia lunata*. *Current Science, 82*, 1457–1463.

Béjà, O., Aravind, L., Koonin, E. V., Suzuki, M. T., Hadd, A., & Nguyen, L. P., et al. (2000a). Bacterial rhodopsin: Evidence for a new type of phototrophy in the sea. *Science, 289*, 1902–1906.

Béjà, O., Spudich, E. N., Spudich, J. L., Leclerc, M., & DeLong, E. F. (2001). Proteorhodopsin phototrophy in the ocean. *Nature, 411*, 786–789.

Béjà, O., Suzuki, M. T., Koonin, E. V., Aravind, L., Hadd, A., & Nguyen, L. P., et al. (2000b). Construction and analysis of bacterial artificial chromosome libraries from a marine microbial assemblage. *Environmental Microbiology, 2*, 516–529.

Berry, A. E., Chiocchini, C., Selby, T., Sosio, M., & Wellington, E. M. H. (2003). Isolation of high molecular weight DNA from soil for cloning into BAC vectors. *FEMS Microbiology Letters, 223*, 15–20.

Blaha, D., Sanguin, H., Robe, P., Nalin, R., Bally, R., & Moenne-Loccoz, Y. (2005). Physical organization of phytobeneficial genes *nifH* and *ipdC* in the plant growth-promoting rhizobacterium *Azospirillum lipoferum* 4V(I). *FEMS Microbiology Letters, 244*, 157–163.

Bloemberg, G. V., & Lugtenberg, B. J. J. (2001). Molecular basis of plant growth promotion and biocontrol by rhizobacteria. *Current Opinion in Plant Biology, 4*, 343–350.

Brady, S. F., & Clardy, J. (2000). Long-chain *n*-acyl amino acid antibiotics isolated from heterologously expressed environmental DNA. *Journal of the American Chemical Society, 122*, 12903–12904.

Breitbart, M., Salamon, P., Andresen, B., Mahaffy, J. M., Segall, A. M., & Mead, D., et al. (2002). Genomic analysis of uncultured marine viral communities. *Proceedings of the National Academy of Sciences USA, 99*, 14250–14255.

Bric, J. M., Bostock, R. M., & Silverstone, S. E. (1991). Rapid in situ assay for indoleacetic-acid production by bacteria immobilized on a nitrocellulose membrane. *Applied and Environmental Microbiology, 57*, 535–538.

Buckley, D. H., & Schmidt, T. M. (2003). Diversity and dynamics of microbial communities in soils from agro-ecosystems. *Environmental Microbiology, 5*, 441–452.

Chen, K., & Pachter, L. (2005). Bioinformatics for whole-genome shotgun sequencing of microbial communities. *PLoS Computational Biology, 1*, 106–112.

Chin-A-Woeng, T. F. C., Bloemberg, G. V., van der Bij, A. J., van der Drift, K. M. G. F., Schripsema, J., & Kroon, B., et al. (1998). Biocontrol by phenazine-1-carboxamide-producing *Pseudomonas chlororaphis* PCL1391 of tomato root rot caused by *Fusarium oxysporum* f. sp. *radicis-lycopersici*. *Molecular Plant–Microbe Interactions, 11*, 1069–1077.

Coeffet-Le Gal, M. F., Thurston, L., Rich, P., Miao, V., & Baltz, R. H. (2006). Complementation of daptomycin *dptA* and *dptD* deletion mutations in trans and production of hybrid lipopeptide antibiotics. *Microbiology, 152*, 2993–3001.

Costa, R., Gotz, M., Mrotzek, N., Lottmann, J., Berg, G., & Smalla, K. (2006a). Effects of site and plant species on rhizosphere community structure as revealed by molecular analysis of microbial guilds. *FEMS Microbiology Ecology, 56*, 236–249.

Costa, R., Salles, J. F., Berg, G., & Smalla, K. (2006b). Cultivation-independent analysis of *Pseudomonas* species in soil and in the rhizosphere of field-grown *Verticillium dahliae* host plants. *Environmental Microbiology, 8*, 2136–2149.

Courtois, S., Cappellano, C. M., Ball, M., Francou, F. X., Normand, P., & Helynck, G., et al. (2003). Recombinant environmental libraries provide access to microbial diversity for drug discovery from natural products. *Applied and Environmental Microbiology, 69*, 49–55.

Daniel, R. (2005). The metagenomics of soil. *Nature Reviews Microbiology, 3*, 470–478.

Delcher, A. L., Harmon, D., Kasif, S., White, O., & Salzberg, S. L. (1999). Improved microbial gene identification with GLIMMER. *Nucleic Acids Research, 27*, 4636–4641.

DeLong, E. F. (2004). Microbial population genomics and ecology: The road ahead. *Environmental Microbiology, 6*, 875–878.

DeLong, E. F., Preston, C. M., Mincer, T., Rich, V., Hallam, S. J., & Frigaard, N. U., et al. (2006). Community genomics among stratified microbial assemblages in the ocean's interior. *Science, 311*, 496–503.

Deutschbauer, A. M., Chivian, D., & Arkin, A. P. (2006). Genomics for environmental microbiology. *Current Opinion in Biotechnology, 17*, 229–235.

Devescovi, G., Aguilar, C., Majolini, M. B., Marugg, J., Weisbeek, P., & Venturi, V. (2001). A siderophore peptide synthetase gene from plant growth-promoting *Pseudomonas putida* WCS358. *Systematic and Applied Microbiology, 24*, 321–330.

Ding, Y., Wang, J., Liu, Y., & Chen, S. (2005). Isolation and identification of nitrogen-fixing bacilli from plant rhizospheres in Beijing region. *Journal of Applied Microbiology, 99*, 1271–1281.

Downing, K., & Thomson, J. A. (2000). Introduction of the *Serratia marcescens chiA* gene into an endophytic *Pseudomonas fluorescens* for the biocontrol of phytopathogenic fungi. *Canadian Journal of Microbiology, 46*, 363– 369.

Dumont, M. G., Radajewski, S. M., Miguez, C. B., McDonald, I. R., & Murrell, J. C. (2006). Identification of a complete methane monooxygenase operon from soil by combining stable isotope probing and metagenomic analysis. *Environmental Microbiology, 8*, 1240–1250.

Edwards, R. A., Rodriguez-Brito, B., Wegley, L., Haynes, M., Breitbart, M., & Peterson, D. M., et al. (2006). Using pyrosequencing to shed light on deep mine microbial ecology. *BMC Genomics, 7*, 57.

Edwards, R. A., & Rohwer, F. (2005). Viral metagenomics. *Nature Reviews Microbiology, 3*, 504–510.

Emmert, E. A. B., Klimowicz, A. K., Thomas, M. G., & Handelsman, J. (2004). Genetics of Zwittermicin A production by *Bacillus cereus*. *Applied and Environmental Microbiology, 70*, 104–113.

Entcheva, P., Liebl, W., Johann, A., Hartsch, T., & Streit, W. R. (2001). Direct cloning from enrichment cultures, a reliable strategy for isolation of complete operons and genes from microbial consortia. *Applied and Environmental Microbiology, 67*, 89–99.

Erkel, C., Kube, M., Reinhardt, R., & Liesack, W. (2006). Genome of Rice Cluster I archaea - the key methane producers in the rice rhizosphere. *Science, 313*, 370–372.

Fould, S., Dieng, A. L., Davies, K. G., Normand, P., & Mateille, T. (2001). Immunological quantification of the nematode parasitic bacterium *Pasteuria penetrans* in soil. *FEMS Microbiology Ecology, 37*, 187–195.

Friedrich, M. W. (2006). Stable-isotope probing of DNA: Insights into the function of uncultivated microorganisms from isotopically labeled metagenomes. *Current Opinion in Biotechnology, 17*, 59–66.

Fuhrman, J. A., Comeau, D. E., Hagstrom, A., & Chan, A. M. (1988). Extraction from natural planktonic microorganisms of DNA suitable for molecular biological studies. *Applied and Environmental Microbiology, 54*, 1426–1429.

Gabor, E. M., Alkema, W. B. L., & Janssen, D. B. (2004a). Quantifying the accessibility of the metagenome by random expression cloning techniques. *Environmental Microbiology, 6*, 879–886.

Gabor, E. M., de Vries, E. J., & Janssen, D. B. (2004b). Construction, characterization, and use of small-insert gene banks of DNA isolated from soil and enrichment cultures for the recovery of novel amidases. *Environmental Microbiology, 6*, 948–958.

Galbraith, E. A., Antonopoulos, D. A., & White, B. A. (2004). Suppressive subtractive hybridization as a tool for identifying genetic diversity in an environmental metagenome: The rumen as a model. *Environmental Microbiology, 6*, 928–937.

Gill, S. R., Pop, M., DeBoy, R. T., Eckburg, P. B., Turnbaugh, P. J., & Samuel, B. S., et al. (2006). Metagenomic analysis of the human distal gut microbiome. *Science, 312*, 1355–1359.

Girard, G., Barends, S., Rigali, S., van Rij, E. T., Lugtenberg, B. J. J., & Bloemberg, G. V. (2006). Pip, a novel activator of phenazine biosynthesis in *Pseudomonas chlororaphis* PCL1391. *Journal of Bacteriology, 188*, 8283–8293.

Glick, B. R., & Bashan, Y. (1997). Genetic manipulation of plant growth-promoting bacteria to enhance biocontrol of phytopathogens. *Biotechnology Advances, 15*, 353–378.

Glick, B. R., Penrose, D. M., & Li, J. P. (1998). A model for the lowering of plant ethylene concentrations by plant growth-promoting bacteria. *Journal of Theoretical Biology, 190*, 63–68.

Gohel, V., Megha, C., Vyas, P., & Chhatpar, H. S. (2004). Strain improvement of chitinolytic enzyme producing isolate *Pantoea dispersa* for enhancing its biocontrol potential against fungal plant pathogens. *Annals of Microbiology, 54*, 503–515.

Grant, S., Grant, W. D., Cowan, D. A., Jones, B. E., Ma, Y. H., & Ventosa, A., et al. (2006). Identification of eukaryotic open reading frames in metagenomic cDNA libraries made from environmental samples. *Applied and Environmental Microbiology, 72*, 135–143.

Gray, E. J., & Smith, D. L. (2005). Intracellular and extracellular PGPR: Commonalities and distinctions in the plant–bacterium signaling processes. *Soil Biology & Biochemistry, 37*, 395–412.

Green, B. D., & Keller, M. (2006). Capturing the uncultivated majority. *Current Opinion in Biotechnology, 17*, 236–240.

Gremion, F., Chatzinotas, A., & Harms, H. (2003). Comparative 16S rDNA and 16S rRNA sequence analysis indicates that *Actinobacteria* might be a dominant part of the metabolically active bacteria in heavy metal-contaminated bulk and rhizosphere soil. *Environmental Microbiology, 5*, 896– 907.

Hallam, S. J., Putnam, N., Preston, C. M., Detter, J. C., Rokhsar, D., & Richardson, P. M., et al. (2004). Reverse methanogenesis: Testing the hypothesis with environmental genomics. *Science, 305*, 1457–1462.

Handelsman, J. (2004). Metagenomics: Application of genomics to uncultured microorganisms. *Microbiology and Molecular Biology Reviews, 68*, 669–685.

Handelsman, J. (2005). Metagenomics or megagenomics? *Nature Reviews Microbiology, 3*, 457–458.

Handelsman, J., Rondon, M. R., Brady, S. F., Clardy, J., & Goodman, R. M. (1998). Molecular biological access to the chemistry of unknown soil microbes: A new frontier for natural products. *Chemistry & Biology, 5*, R245–R249.

Hashidoko, Y., Tada, M., Osaki, M., & Tahara, S. (2002). Soft gel medium solidified with gellan gum for preliminary screening for root-associating, free-living nitrogen-fixing bacteria inhabiting the rhizoplane of plants. *Bioscience Biotechnology and Biochemistry, 66*, 2259–2263.

Healy, F. G., Ray, R. M., Aldrich, H. C., Wilkie, A. C., Ingram, L. O., & Shanmugam, K. T. (1995). Direct isolation of functional genes encoding cellulases from the microbial consortia in a thermophilic, anaerobic digester maintained on lignocellulose. *Applied Microbiology and Biotechnology, 43*, 667–674.

Henne, A., Daniel, R., Schmitz, R. A., & Gottschalk, G. (1999). Construction of environmental DNA libraries in *Escherichia coli* and screening for the presence of genes conferring utilization of 4-hydroxybutyrate. *Applied and Environmental Microbiology, 65*, 3901–3907.

Holguin, G., & Glick, B. R. (2003). Transformation of *Azospirillum brasilense* Cd with an ACC deaminase gene from *Enterobacter cloacae* UW4 fused to the Tet(r) gene promoter improves its fitness and plant growth promoting ability. *Microbial Ecology, 46*, 122–133.

Hurek, T., Handley, L. L., Reinhold-Hurek, B., & Piche, Y. (2002). *Azoarcus*. grass endophytes contribute fixed nitrogen to the plant in an unculturable state. *Molecular Plant–Microbe Interactions, 15*, 233–242.

Hurt, R. A., Qiu, X. Y., Wu, L. Y., Roh, Y., Palumbo, A. V., & Tiedje, J. M., et al. (2001). Simultaneous recovery of RNA and DNA from soils and sediments. *Applied and Environmental Microbiology, 67*, 4495–4503.

Jacobsen, C. S. (1995). Microscale detection of specific bacterial-DNA in soil with a magnetic capture-hybridization and PCR amplification assay. *Applied and Environmental Microbiology, 61*, 3347–3352.

Jacobsen, C. S. (2004). Purification of microbial genes from soil and rhizosphere by magnetic capture hybridization and subsequent amplification of target genes by PCR. In: G. A. Kowalchuk, F. J. De Bruijn, I. M. Head, A. D. L. Akkermans,

J. D. Van Elsas (Eds.), *Molecular microbial ecology manual* (pp. 181–188). Dordrecht, The Netherlands: Kluwer.

Jeong, H., Kim, J. F., Park, Y. K., Kim, S. B., Kim, C., & Park, S. H. (2006). Genome snapshot of *Paenibacillus polymyxa* ATCC 842. *Journal of Microbiology and Biotechnology, 16*, 1650–1655.

Jiao, J. Y., Wang, H. X., Zeng, Y., & Shen, Y. M. (2006). Enrichment for microbes living in association with plant tissues. *Journal of Applied Microbiology, 100*, 830–837.

Joseph, S. J., Hugenholtz, P., Sangwan, P., Osburne, C. A., & Janssen, P. H. (2003). Laboratory cultivation of widespread and previously uncultured soil bacteria. *Applied and Environmental Microbiology, 69*, 7210–7215.

Juraeva, D., George, E., Davranov, K., & Ruppel, S. (2006). Detection and quantification of the *nifH* gene in shoot and root of cucumber plants. *Canadian Journal of Microbiology, 52*, 731–739.

Kaeberlein, T., Lewis, K., & Epstein, S. S. (2002). Isolating 'uncultivable' microorganisms in pure culture by a simulated natural environment. *Science, 296*, 1127–1129.

Kim, J., Kim, E., Kang, Y., Choi, O., Park, C. S., & Hwang, I. (2006). Molecular characterization of biosynthetic genes of an antifungal compound produced by *Pseudomonas fluorescens* MC07. *Journal of Microbiology and Biotechnology, 16*, 450–456.

Kloepper, J. W., & Schroth, M. N. (1978). Plant growth-promoting rhizobacteria in radish. In: Proceedings of the 4th International Conference on Plant Pathogenic Bacteria (pp. 879–882), INRA, Angers, France.

Knietsch, A., Bowien, S., Whited, G., Gottschalk, G., & Daniel, R. (2003a). Identification and characterization of coenzyme B-12-dependent glycerol dehydratase- and diol dehydratase-encoding genes from metagenomic DNA libraries derived from enrichment cultures. *Applied and Environmental Microbiology, 69*, 3048–3060.

Knietsch, A., Waschkowitz, T., Bowien, S., Henne, A., & Daniel, R. (2003b). Construction and screening of metagenomic libraries derived from enrichment cultures: Generation of a gene bank for genes conferring alcohol oxidoreductase activity on *Escherichia coli. Applied and Environmental Microbiology, 69*, 1408–1416.

Kobayashi, D. Y., Reedy, R. M., Bick, J., & Oudemans, P. V. (2002). Characterization of a chitinase gene from *Stenotrophomonas maltophilia* strain 34S1 and its involvement in biological control. *Applied and Environmental Microbiology, 68*, 1047–1054.

Kowalchuk, G. A., de Bruijn, F. J., Head, I. M., Akkermans, A. D. L., & van Elsas, J. D. (2004). *Molecular microbial ecology manual*. Dordrecht, The Netherlands: Kluwer.

Kowalchuk, G. A., Speksnijder, A. G. C. L., Zhang, K., Goodman, R. M., & van Veen, J. A. (2007). Finding the needles in the metagenome haystack. *Microbial Ecology, 53*, 475–485.

Kozdroj, J., & van Elsas, J. D. (2000). Application of polymerase chain reaction-denaturing gradient gel electrophoresis for comparison of direct and indirect extraction methods of soil DNA used for microbial community fingerprinting. *Biology and Fertility of Soils, 31*, 372–378.

Lee, E. T., Lim, S. K., Nam, D. H., Khang, Y. H., & Kim, S. D. (2003). Pyoverdin(2112) of *Pseudomonas fluorescens* 2112 inhibits *Phytophthora capsici*, a red-pepper blight-causing fungus. *Journal of Microbiology and Biotechnology, 13*, 415–421.

Leveau, J. H. J., Gerards, S., de Boer, W., & van Veen, J. A. (2004). Phylogeny-function analysis of (meta)genomic libraries: Screening for expression of ribosomal RNA genes by large-insert library fluorescent in situ hybridization (LIL-FISH). *Environmental Microbiology, 6*, 990–998.

Leveau, J. H. J., Gerards, S., Fritsche, K., Zondag, G., & van Veen, J. A. (2006). Genomic flank-sequencing of plasposon insertion sites for rapid identification of functional genes. *Journal of Microbiological Methods, 66*, 276–285.

Leveau, J. H. J., & Lindow, S. E. (2005). Utilization of the plant hormone indole-3-acetic acid for growth by *Pseudomonas putida* strain 1290. *Applied and Environmental Microbiology, 71*, 2365–2371.

Li, Y. G., Wexler, M., Richardson, D. J., Bond, P. L., & Johnston, A. W. B. (2005). Screening a wide host-range, waste-water metagenomic library in tryptophan auxotrophs of *Rhizobium leguminosarum* and of *Escherichia coli* reveals different classes of cloned *trp* genes. *Environmental Microbiology, 7*, 1927–1936.

Liles, M. R., Manske, B. F., Bintrim, S. B., Handelsman, J., & Goodman, R. M. (2003). A census of rRNA genes and linked genomic sequences within a soil metagenomic library. *Applied and Environmental Microbiology, 69*, 2684–2691.

Liles, M. R., Williamson, L. L., Handelsman, J., & Goodman, R. M. (2004). Isolation of high molecular weight genomic DNA from soil bacteria for genomic library construction. In G. A. Kowalchuk, F. J. De Bruijn, I. M. Head, A. D. L. Akkermans, & J. D. Van Elsas (Eds.) *Molecular microbial ecology manual* (pp. 839–849)). Dordrecht, The Netherlands: Kluwer.

Lindahl, V., & Bakken, L. R. (1995). Evaluation of methods for extraction of bacteria from soil. *FEMS Microbiology Ecology, 16*, 135–142.

Lorenz, P., & Eck, J. (2005). Metagenomics and industrial applications. *Nature Reviews Microbiology, 3*, 510–516.

Lucy, M., Reed, E., & Glick, B. R. (2004). Applications of free living plant growth-promoting rhizobacteria. *Antonie van Leeuwenhoek, 86*, 1–25.

Luna, G. M., Dell'Anno, A., & Danovaro, R. (2006). DNA extraction procedure: A critical issue for bacterial diversity assessment in marine sediments. *Environmental Microbiology, 8*, 308–320.

Majernik, A., Gottschalk, G., & Daniel, R. (2001). Screening of environmental DNA libraries for the presence of genes conferring $Na^{+}(Li^{+})/H^{+}$ antiporter activity on *Escherichia coli*: Characterization of the recovered genes and the corresponding gene products. *Journal of Bacteriology, 183*, 6645–6653.

Marchesi, J. R., & Weightman, A. J. (2003). Comparing the dehalogenase gene pool in cultivated . a. -halocarboxylic acid-degrading bacteria with the environmental metagene pool. *Applied and Environmental Microbiology, 69*, 4375–4382.

Margulies, M., Egholm, M., Altman, W. E., Attiya, S., Bader, J. S., & Bemben, L. A., et al. (2005). Genome sequencing in microfabricated high-density picolitre reactors. *Nature, 437*, 376–380.

Mark, G. L., Dow, J. M., Kiely, P. D., Higgins, H., Haynes, J., & Baysse, C., et al. (2005). Transcriptome profiling of bacterial responses to root exudates identifies genes involved in microbe–plant interactions. *Proceedings of the National Academy of Sciences USA*, *102*, 17454–17459.

Martinez, A., Kolvek, S. J., Yip, C. L. T., Hopke, J., Brown, K. A., & MacNeil, I. A., et al. (2004). Genetically modified bacterial strains and novel bacterial artificial chromosome shuttle vectors for constructing environmental libraries and detecting heterologous natural products in multiple expression hosts. *Applied and Environmental Microbiology*, *70*, 2452–2463.

Meyer, F., Goesmann, A., McHardy, A. C., Bartels, D., Bekel, T., & Clausen, J., et al. (2003). GenDB – An open source genome annotation system for prokaryote genomes. *Nucleic Acids Research*, *31*, 2187–2195.

Miller, D. N., Bryant, J. E., Madsen, E. L., & Ghiorse, W. C. (1999). Evaluation and optimization of DNA extraction and purification procedures for soil and sediment samples. *Applied and Environmental Microbiology*, *65*, 4715–4724.

Millner, P. D., & Wright, S. F. (2002). Tools for support of ecological research on arbuscular mycorrhizal fungi. *Symbiosis*, *33*, 101–123.

Mills, H. J., Martinez, R. J., Story, S., & Sobecky, P. A. (2004). Identification of members of the metabolically active microbial populations associated with *Beggiatoa* species mat communities form Gulf of Mexico cold-seep sediments. *Applied and Environmental Microbiology*, *70*, 5447–5458.

Nesbø, C. L., Boucher, Y., Dlutek, M., & Doolittle, W. F. (2005). Lateral gene transfer and phylogenetic assignment of environmental fosmid clones. *Environmental Microbiology*, *7*, 2011–2026.

Nocker, A., & Camper, A. K. (2006). Selective removal of DNA from dead cells of mixed bacterial communities by use of ethidium monoazide. *Applied and Environmental Microbiology*, *72*, 1997–2004.

Noguchi, H., Park, J., & Takagi, T. (2006). MetaGene: Prokaryotic gene finding from environmental genome shotgun sequences. *Nucleic Acids Research*, *34*, 5623–5630.

Omer, Z. S., Tombolini, R., Broberg, A., & Gerhardson, B. (2004). Indole-3-acetic acid production by pink-pigmented facultative methylotrophic bacteria. *Plant Growth Regulation*, *43*, 93–96.

Oremland, R. S., Capone, D. G., Stolz, J. F., & Fuhrman, J. (2005). Whither or wither geomicrobiology in the era of "community metagenomics". *Nature Reviews Microbiology*, *3*, 572–578.

Poretsky, R. S., Bano, N., Buchan, A., LeCleir, G., Kleikemper, J., & Pickering, M., et al. (2005). Analysis of microbial gene transcripts in environmental samples. *Applied and Environmental Microbiology*, *71*, 4121–4126.

Quaiser, A., Ochsenreiter, T., Klenk, H. P., Kletzin, A., Treusch, A. H., & Meurer, G., et al. (2002). First insight into the genome of an uncultivated crenarchaeote from soil. *Environmental Microbiology*, *4*, 603–611.

Quaiser, A., Ochsenreiter, T., Lanz, C., Schuster, S. C., Treusch, A. H., & Eck, J., et al. (2003). Acidobacteria form a coherent but highly diverse group within the bacterial domain: Evidence from environmental genomics. *Molecular Microbiology*, *50*, 563–575.

Raaijmakers, J. M., de Bruijn, I., & de Kock, M. J. D. (2006). Cyclic lipopeptide production by plant-associated *Pseudomonas* spp.: Diversity, activity, biosynthesis, and regulation. *Molecular Plant–Microbe Interactions*, *19*, 699–710.

Radwan, T., Mohamed, Z. K., & Reis, V. M. (2002). Production of indole-3-acetic acid by different strains of *Azospirillum* and *Herbaspirillum* spp. *Symbiosis*, *32*, 39–53.

Rajendran, N., Rompf, A., Marahiel, M. A., & Jahn, D. (1998). Isolation and characterization of a transposon mutant of *Pseudomonas fluorescens* AU63 deficient in antifungal activity against *Pythium ultimum*. *Letters in Applied Microbiology*, *27*, 152–157.

Ram, R. J., VerBerkmoes, N. C., Thelen, M. P., Tyson, G. W., Baker, B. J., & Blake, R. C., et al. (2005). Community proteomics of a natural microbial biofilm. *Science*, *308*, 1915–1920.

Rangarajan, S., Saleena, L. M., Vasudevan, P., & Nair, S. (2003). Biological suppression of rice diseases by *Pseudomonas* spp. under saline soil conditions. *Plant and Soil*, *251*, 73–82.

Riesenfeld, C. S., Goodman, R. M., & Handelsman, J. (2004a). Uncultured soil bacteria are a reservoir of new antibiotic resistance genes. *Environmental Microbiology*, *6*, 981–989.

Riesenfeld, C. S., Schloss, P. D., & Handelsman, J. (2004b). Metagenomics: Genomic analysis of microbial communities. *Annual Reviews in Genetics*, *38*, 525–552.

Rodriguez, H., Gonzalez, T., & Selman, G. (2000). Expression of a mineral phosphate solubilizing gene from *Erwinia herbicola* in two rhizobacterial strains. *Journal of Biotechnology*, *84*, 155–161.

Rodriguez-Brito, B., Rohwer, F., & Edwards, R. A. (2006). An application of statistics to comparative metagenomics. *BMC Bioinformatics*, *7*, 162.

Rondon, M. R., August, P. R., Bettermann, A. D., Brady, S. F., Grossman, T. H., & Liles, M. R., et al. (2000). Cloning the soil metagenome: A strategy for accessing the genetic and functional diversity of uncultured microorganisms. *Applied and Environmental Microbiology*, *66*, 2541–2547.

Rutherford, K., Parkhill, J., Crook, J., Horsnell, T., Rice, P., & Rajandream, M. A. (2000). Artemis: Sequence visualization and annotation. *Bioinformatics*, *16*, 944–945.

Sanger, F., Nicklen, S., & Coulson, A. R. (1977). DNA Sequencing with chain-terminating inhibitors. *Proceedings of the National Academy of Sciences of the United States of America*, *74*, 5463–5467.

Sanguin, H., Remenant, B., Dechesne, A., Thioulouse, J., Vogel, T. M., & Nesme, X., et al. (2006). Potential of a 16S rRNA-based taxonomic microarray for analyzing the rhizosphere effects of maize on *Agrobacterium* spp. and bacterial communities. *Applied and Environmental Microbiology*, *72*, 4302–4312.

Sato, M., Watanabe, K., Yazawa, M., Takikawa, Y., & Nishiyama, K. (1997). Detection of new ethylene-producing bacteria, *Pseudomonas syringae* pvs. *cannabina* and *sesami*, by PCR amplification of genes for the ethylene-forming enzyme. *Phytopathology*, *87*, 1192–1196.

Schloss, P. D., & Handelsman, J. (2005). Metagenomics for studying unculturable microorganisms: Cutting the Gordian knot. *Genome Biology*, *6*, 229.

Schmeisser, C., Steele, H., & Streit, W. R. (2007). Metagenomics, biotechnology with non-culturable microbes. *Applied Microbiology and Biotechnology*, *75*, 955–962.

Schmeisser, C., Stockigt, C., Raasch, C., Wingender, J., Timmis, K. N., & Wenderoth, D. F., et al. (2003). Metagenome survey of biofilms in drinking-water networks. *Applied and Environmental Microbiology, 69*, 7298–7309.

Schmidt, T. M., Delong, E. F., & Pace, N. R. (1991). Analysis of a marine picoplankton community by 16S ribosomal-RNA gene cloning and sequencing. *Journal of Bacteriology, 173*, 4371–4378.

Schwarz, S., Waschkowitz, T., & Daniel, R. (2006). Enhancement of gene detection frequencies by combining DNA-based stable-isotope probing with the construction of metagenomic DNA libraries. *World Journal of Microbiology & Biotechnology, 22*, 363–367.

Shah, S., Li, J. P., Moffatt, B. A., & Glick, B. R. (1998). Isolation and characterization of ACC deaminase genes from two different plant growth-promoting rhizobacteria. *Canadian Journal of Microbiology, 44*, 833–843.

Shaharoona, B., Arshad, M., & Zahir, Z. A. (2006). Effect of plant growth-promoting rhizobacteria containing ACC-deaminase on maize (*Zea mays* L.) growth under axenic conditions and on nodulation in mung bean (*Vigna radiata* L.). *Letters in Applied Microbiology, 42*, 155–159.

Sliwinski, M. K., & Goodman, R. M. (2004). Comparison of crenarchaeal consortia inhabiting the rhizosphere of diverse terrestrial plants with those in bulk soil in native environments. *Applied and Environmental Microbiology, 70*, 1821–1826.

Sreenivasan, A. (2001). Earth's anonymous inhabitants. In: Science Notes, an online magazine produced by the graduate students in the Science Communication Program at the University of California, Santa Cruz (URL to the article: http://scicom.ucsc.edu/SciNotes/0101/bacteria.html).

Staunton, J., & Weissman, K. J. (2001). Polyketide biosynthesis: A millennium review. *Natural Product Reports, 18*, 380–416.

Stevenson, B. S., Eichorst, S. A., Wertz, J. T., Schmidt, T. M., & Breznak, J. A. (2004). New strategies for cultivation and detection of previously uncultured microbes. *Applied and Environmental Microbiology, 70*, 4748–4755.

Streit, W. R., & Schmitz, R. A. (2004). Metagenomics – The key to the uncultured microbes. *Current Opinion in Microbiology, 7*, 492–498.

Tebbe, C. C., & Vahjen, W. (1993). Interference of humic acids and DNA extracted directly from soil in detection and transformation of recombinant-DNA from bacteria and a yeast. *Applied and Environmental Microbiology, 59*, 2657–2665.

Teeling, H., Waldmann, J., Lombardot, T., Bauer, M., & Glockner, F. O. (2004). TETRA: A web-service and a stand-alone program for the analysis and comparison of tetranucleotide usage patterns in DNA sequences. *BMC Bioinformatics, 5*, 163.

Tejera, N., Lluch, C., Martinez-Toledo, M. V., & Gonzalez-Lopez, J. (2005). Isolation and characterization of *Azotobacter* and *Azospirillum* strains from the sugarcane rhizosphere. *Plant and Soil, 270*, 223–232.

Timms-Wilson, T. M., Kilshaw, K., & Bailey, M. J. (2004). Risk assessment for engineered bacteria used in biocontrol of fungal disease in agricultural crops. *Plant and Soil, 266*, 57–67.

Timmusk, S., Nicander, B., Granhall, U., & Tillberg, E. (1999). Cytokinin production by *Paenibacillus polymyxa*. *Soil Biology & Biochemistry, 31*, 1847–1852.

Tringe, S. G., & Rubin, E. M. (2005). Metagenomics: DNA sequencing of environmental samples. *Nature Reviews Genetics, 6*, 805–814.

Tringe, S. G., von Mering, C., Kobayashi, A., Salamov, A. A., Chen, K., & Chang, H. W., et al. (2005). Comparative metagenomics of microbial communities. *Science, 308*, 554–557.

Tyson, G. W., Chapman, J., Hugenholtz, P., Allen, E. E., Ram, R. J., & Richardson, P. M., et al. (2004). Community structure and metabolism through reconstruction of microbial genomes from the environment. *Nature, 428*, 37–43.

Tyson, G. W., Lo, I., Baker, B. J., Allen, E. E., Hugenholtz, P., & Banfield, J. F. (2005). Genome-directed isolation of the key nitrogen fixer *Leptospirillum ferrodiazotrophum* sp nov from an acidophilic microbial community. *Applied and Environmental Microbiology, 71*, 6319–6324.

Uchiyama, T., Abe, T., Ikemura, T., & Watanabe, K. (2005). Substrate-induced gene-expression screening of environmental metagenome libraries for isolation of catabolic genes. *Nature Biotechnology, 23*, 88–93.

Venter, J. C., Adams, M. D., Myers, E. W., Li, P. W., Mural, R. J., & Sutton, G. G., et al. (2001). The sequence of the human genome. *Science, 291*, 1304–1351.

Venter, J. C., Remington, K., Heidelberg, J. F., Halpern, A. L., Rusch, D., & Eisen, J. A., et al. (2004). Environmental genome shotgun sequencing of the Sargasso Sea. *Science, 304*, 66–74.

Vessey, J. K. (2003). Plant growth-promoting rhizobacteria as biofertilizers. *Plant and Soil, 255*, 571–586.

Wagner, M., & Horn, M. (2006). The Planctomycetes, Verrucomicrobia, Chlamydiae and sister phyla comprise a superphylum with biotechnological and medical relevance. *Current Opinion in Biotechnology, 17*, 241–249.

Wang, K., Conn, K., & Lazarovits, G. (2006). Involvement of quinolinate phosphoribosyl transferase in promotion of potato growth by a *Burkholderia* strain. *Applied and Environmental Microbiology, 72*, 760–768.

Wang, G. Y. S., Graziani, E., Waters, B., Pan, W., Li, X., & McDermott, J., et al. (2000). Novel natural products from soil DNA libraries in a Streptomycete host. *Organic Letters, 2*, 2401–2404.

Wang, Y. Q., Ohara, Y., Nakayashiki, H., Tosa, Y., & Mayama, S. (2005). Microarray analysis of the gene expression profile induced by the endophytic plant growth-promoting rhizobacteria, *Pseudomonas fluorescens* FPT9601-T5 in *Arabidopsis*. *Molecular Plant–Microbe Interactions, 18*, 385–396.

Ward, N. (2006). New directions and interactions in metagenomics research. *FEMS Microbiology Ecology, 55*, 331–338.

Weller, D. M., Raaijmakers, J. M., Gardener, B. B. M., & Thomashow, L. S. (2002). Microbial populations responsible for specific soil suppressiveness to plant pathogens. *Annual Review of Phytopathology, 40*, 309–348.

Wenzel, S. C., & Müller, R. (2005). Recent developments towards the heterologous expression of complex bacterial natural product biosynthetic pathways. *Current Opinion in Biotechnology, 16*, 594–606.

Whitaker, R. J., & Banfield, J. F. (2006). Population genomics in natural microbial communities. *Trends in Ecology & Evolution, 21*, 508–516.

Williamson, L. L., Borlee, B. R., Schloss, P. D., Guan, C. H., Allen, H. K., & Handelsman, J. (2005). Intracellular screen to identify metagenomic clones that induce or inhibit a quorum-sensing biosensor. *Applied and Environmental Microbiology, 71*, 6335–6344.

Woyke, T., Teeling, H., Ivanova, N. N., Huntemann, M., Richter, M., & Gloeckner, F. O., et al. (2006). Symbiosis insights through metagenomic analysis of a microbial consortium. *Nature, 443*, 950–955.

Xu, J. P. (2006). Microbial ecology in the age of genomics and metagenomics: Concepts, tools, and recent advances. *Molecular Ecology, 15*, 1713–1731.

Zengler, K., Toledo, G., Rappe, M., Elkins, J., Mathur, E. J., & Short, J. M., et al. (2002). Cultivating the uncultured. *Proceedings of the National Academy of Sciences of the United States of America, 99*, 15681–15686.

Zhou, J. Z., Bruns, M. A., & Tiedje, J. M. (1996). DNA recovery from soils of diverse composition. *Applied and Environmental Microbiology, 62*, 316–322.

Eur J Plant Pathol (2007) 119:301–309
DOI 10.1007/s10658-007-9171-3

REVIEW PAPER

Microscopic analysis of plant–bacterium interactions using auto fluorescent proteins

Guido V. Bloemberg

Received: 22 January 2007 / Accepted: 7 May 2007 / Published online: 12 June 2007

Abstract Plant growth promoting rhizobacteria (PGPR) include bacteria that fix nitrogen (e.g., *Rhizobiaceae*, *Herbaspirillum*, *Azoarcus*), produce phytohormones (e.g., *Azospirillum*) and provide protection against fungal and/or bacterial pathogens (e.g., *Pseudomonas*, *Bacillus*, *Streptomyces*). Interactions between PGPR and plants can be divided into different steps which include initial attraction, attachment, proliferation and colonization e.g., of roots, stem, leaves and flowers. At the genetic level the expression of many bacterial genes are altered during these processes. In addition to the interaction with the plant, PGPR interact and compete with the endogenous microflora, consisting of other bacteria, fungi and/or mycorrhizal fungi. In the case of biocontrol bacterial strains, a direct interaction with the pathogen is often required to suppress the disease. Microscopic analyses of plant growth promoting rhizobacteria (PGPR) in their natural environment and in specific during their interaction(s) with the host plant(s) and/or their target organism(s) is essential for the elucidation of their functioning and the successful application of commercial inoculants. With the discovery and development of auto fluorescent proteins (AFPs) as markers and the development of highly sophisticated fluorescence microscopes such as confocal laser scanning microscopes, a new dimension has been created for studying PGPR in their natural environment. This paper will give a short overview on available tools, the application of AFPs in PGPR research and some future perspectives. Several recent reviews will give the reader an option for further reading (Bloemberg and Lugtenberg 2004; Chalfie and Kain 2005; Larrainzar et al. 2005; Rediers et al. 2005; Bloemberg and Camacho 2006).

G. V. Bloemberg (✉)
Institute of Biology, Leiden University, Wassenaarseweg 64, 2333AL Leiden, The Netherlands
e-mail: g.v.bloemberg@biology.leidenuniv.nl

Keywords PGPR · Auto fluorescent proteins · GFP · Microscopy · Marker proteins · Plant-microbe interactions

A short overview of available auto fluorescent proteins and their properties

The first application of green fluorescent protein (GFP) isolated from the jellyfish *Aequorea victoria* as a reporter (Chalfie et al. 1994) has become a hallmark in modern biology and is used throughout a wide range of different biology and biotechnology research areas including many fields of microbiology and cell biology. AFPs have become the most important reporters providing new tools to mark whole cells, study protein localization, and monitor gene expression and molecule interactions *in vivo* in a non-invasive way thereby preserving the integrity of the cell. Advantages of the use of GFP in comparison with other reporters or dyes is that GFP is present within the cell as a product of gene expression and that the visualization does not require any fixation or preparation protocols, which are time-consuming and might result in artifacts or influence cellular properties. Furthermore it does not require substrates or additional energy such as often is the case in bioluminescence. Additional advantages are that GFP due to its tertiary barrel structure is very stable and can be applied in many different species. In addition, GFP labelled cells can be used for flow cytometry analysis and quantitative analysis by PCR (Utermark and Karlovsky 2006). Disadvantages of GFP are that its structure and fluorescence is dependent on pH and the presence of oxygen (Heim et al. 1994). However, studies on *Rhizobium* tagged with a GFP-derivative showed that GFP was well visualized in bacteroids present in root nodules, which is an oxygen-limiting environment (Gage et al. 1996; Stuurman et al. 2000). Since GFP is not an enzyme it does not have amplifying reporter ability, such as, for example, LacZ. Successful GFP visualization is therefore a balance between sufficient production and prevention of over-production that could cause toxicity or a metabolic burden to the cell resulting in growth inhibition. The latter should be tested for each bacterial species. Cloning of *e-gfp* under control of the *lac* promoter on a high copy number plasmid in *E.coli* results, for instance, in a severe growth inhibition when grown on Luria-Bertani agar medium (unpublished results). After 2 days of incubation white colonies grow out of the bacterial streak representing spontaneous *gfp* mutants with a growth advantage. Another point of consideration before applying GFP is the autofluorescence background or noise from the environment in which the bacteria are to be analysed. For example, sand and other soil particles can severely hamper GFP visualization as well as certain plant structures or organelles such as chloroplasts. Such problems might be solved by using other auto fluorescent proteins with different excitation and emission wavelength spectra.

Modifications of GFP (often by gene shuffling experiments) have resulted in the isolation of mutants that have shifted emission and excitation wavelengths, which give the opportunity to use multiple auto fluorescent proteins in one system in order to differentiate between different cells or to visualize different processes within one cell. In addition, such variants are also brighter and more stable. Important GFP derivatives are Enhanced GFP (EGFP), Enhanced Cyan Fluorescent Protein (ECFP) and Enhanced Yellow Fluorescent Protein (YFP) (Yang et al. 1998; Tsien 1998; Matus 1999; Ellenberg et al. 1999). Blue fluorescent protein (BFP) has also been developed but is less used due to its low brightness. In addition many other *gfp* derivatives have been isolated or developed with enhanced brightness and optimized codon usage for optimal expression in different organisms. A wide range of auto fluorescent proteins with useful background information is available on the website of the Clontech company (http://www.clontech.com/). Although the stability of GFP is very advantageous for tracking studies, it hampers transient gene expression studies. Andersen et al. (1999) have developed a set of GFP derivatives with reduced half-lifetimes by the addition of short amino acid tags to the C-terminus, recognized by specific proteases widely present in bacterial cells, which usually break down partially produced proteins.

Although many efforts were made to isolate a red fluorescent derivative of GFP, this has never been achieved and was bypassed by the discovery and application of Red Fluorescent Protein (RFP or DsRed) isolated from the coral *Discosoma striata* (Matz et al. 1999). Since the *rfp* sequence is not homologous to *gfp* the use of both genes in one cell will not result in unwanted recombinations. An efficient use of DsRed is hampered by its slow maturation due to its tetramerization, which is required for its fluorescent properties, and its toxic properties when overproduced. Recently, several improved DsRed derivatives have been constructed to overcome these problems. One of these new derivatives DsRed.T3_S4T, which maturates faster (Sorensen et al. 2003) was successfully applied in *Pseudomonas* spp. for rhizosphere studies in being brighter and without causing loss of competitive colonization ability (Dandie et al. 2005). A more recent paper by Shaner et al. (2004) reports on the construction of improved monomeric red, orange and yellow fluorescent proteins derived from DsRed, which mature more efficiently, are more tolerant to N-terminal fusions and have an improved photostability. These forms have not been reported for studies of PGPR.

Genetic tools to mark microorganisms with auto fluorescent proteins

A requirement of the application of AFPs is that the genes encoding these proteins have to be transformed into the bacteria. This requirement can prevent the use of GFP as a marker when a certain bacterial species or strain is not genetically accessible with the available transformation protocols. AFP genes are usually delivered on plasmids or transposons (Bloemberg et al. 2004). The advantage of a plasmid is that it is present in multiple copies, which can improve the production of AFPs and does not disrupt host genes by chromosomal integration.

Since antibiotics cannot usually be applied for in vivo studies of PGPR on the plant the development of a set of broad host range plasmids that are stably maintained (in Gram-negative bacteria) without antibiotic pressure has been extremely valuable (Heeb et al. 2000). We have used such plasmids to construct a set of marker plasmids carrying e*gfp* (green), e*cfp* (cyan), e*yfp* (yellow) e*bfp* (blue) and *rfp* (red), which were successfully used for the visualisation of PGPR such as *Pseudomonas* biocontrol strains and *Rhizobium* strains respectively (Stuurman et al. 2000; Bloemberg et al. 2000). The value of these vectors was also shown for the analysis of the improved DsRed form (Dandie et al. 2005). Another point of attention for an efficient marking is the use of an appropriate promoter. Both the use of a *tac* and *lac* promoter were shown successful for the constitutive expression of AFPs in *Pseudomonas* and *Rhizobium* and many other Gram-negative bacterial strains (Bloemberg et al. 1997, 2000) (Fig. 1). During the past 10 years valuable transposon constructs have also become available, that carry mainly *gfp* derivatives for marking and/or gene expression studies (Burlage et al. 1995; Tombolini et al. 1997; Unge et al. 1997; Xi et al. 1999).

Since fungi are frequently part of the endogenous microflora and can even be the direct target for the PGPR effect as in case of biological control, it is of great relevance to tag fungi and study their interaction with the PGPR. However, genetic transformation of fungi is usually more difficult than transformation of bacterial cells, due to the presence of the rigid cell wall and need for stable integration of genetic material in the chromosome. A classic method consists of the preparation of protoplasts, which will subsequently take up genetic material, mostly in the form of plasmids, which will integrate into the chromosome. Consequently, fungal transformants have to be tested for the conservation of phenotypic properties such as morphology, growth and pathogenicity before using these transformants in confrontation studies. Using an optimized protoplast transformation protocol we have tagged *Fusarium oxysporum* f.sp. *radicis lycopersici* with different autofluorescent proteins by co-transformation with two plasmids of which one contained a hygromycin resistance gene and the second an *afp* gene (Lagopodi et al. 2002; Bolwerk et al. 2003). In addition to protoplastation methods, fungi can be transformed by ballistic bombardment and more recently by the transformation ability of *Agrobacterium* (de Groot et al. 1998), which has been shown to be a simple, efficient and successful transformation method for many fungi (Michielse et al. 2005).

Some examples of the use of auto fluorescent proteins for the visualization of PGPR in the plant environment

Initial studies using auto fluorescent proteins (in specific GFP) as markers for PGPR were used for localization studies. Most of these studies showed that PGPR and other microorganisms such as phytopathogenic fungi preferentially colonize the junctions between the root cells (Bloemberg et al. 1997, 2000; Tombolini et al. 1999; Lagopodi et al. 2002; Bolwerk et al. 2003; Gamalero et al. 2005) (Fig. 1). Most steps in the symbiosis process between *Rhizobiacea* and leguminous plants occur inside the root after entrance into the infection thread. The use of GFP has allowed us to visualize the process of attachment, entrance and nodule occupancy in great detail (Gage et al. 1996) making it even possible to determine the growth rate of the cells in the infection thread (Gage et al. 1996). Stuurman et al. (2000) showed that GFP tagged *Rhizobium* bacteroids move in the root nodule. Since the root is sometimes too thick for successful imaging due to loss of light in the deeper root parts, sectioning of the plant material can solve this. This is also used for the study of endophytes as was shown by several publications on *Herbaspirillum* spp. (Elbeltagy et al. 2001) and the pathogen invasion of *Xylella fastidiosa* (Newman et al. 2003). When necessary plant material can also be stored before visualization by fixation with paraformaldehyde, which leaves GFP intact for fluorescent studies (Stuurman et al. 2000; Elbeltagy et al. 2001). Confocal laser scanning microscopy (CLSM) analysis of the colonization behaviour of *afp*-tagged antagonistic strains can also provide important information on the sampling strategy

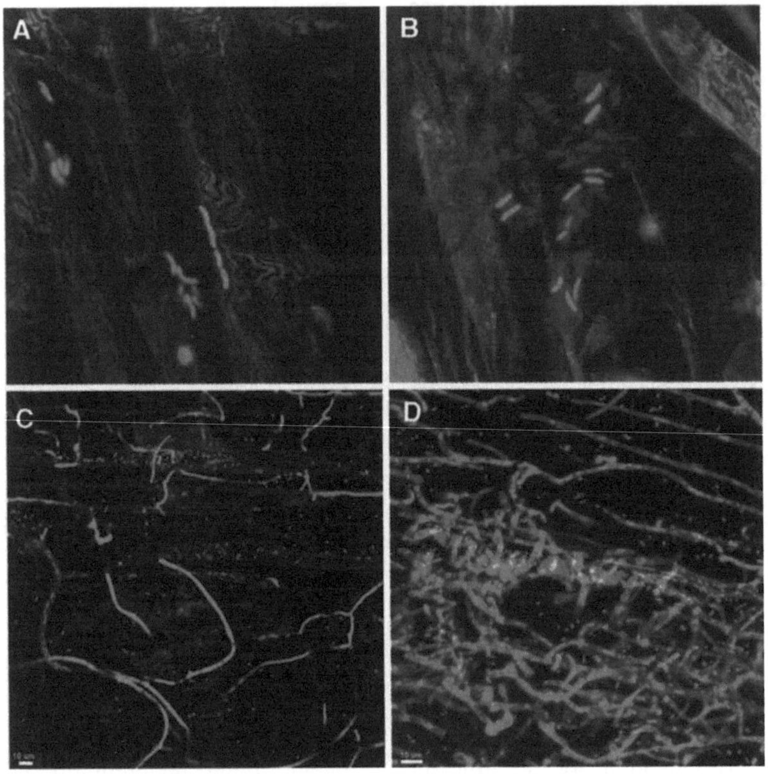

Fig. 1 Confocal laser scanning microscopy (CLSM) analyses of *Pseudomonas* biocontrol strains and the pathogen *Fusarium oxysporum* f.sp. *radicis lycopersici* marker with auto fluorescent proteins in the tomato rhizosphere. Panels (**A**) *P. fluorescens* WCS36 marked with Green Fluorescent Protein (GFP) on the tomato root surface; (**B**) Colonies of *P. fluorescens* WCS365 after inoculation of a mixed culture of GFP and cyan fluorescent protein (CFP) marked derivative strains; (**C** and **D**) Simultaneous imaging of *P. fluorescens* WCS365 marked with red fluorescent protein (DsRed) and *Fusarium oxysporum* f.sp. *radicis lycopserisci* marked with GFP. (images were produced by A. Wijfjes, A. Lagopodi and A. Bolwerk)

required for monitoring inoculant strains and combine these with data on the kinetics of the endogenous microflora (Gotz et al. 2006).

Localization studies are also valuable for the analysis of the properties of mutant strains, for instance for their adhering abilities. For example Biancotto et al. (2001) showed by using mucoid mutant strains of *P. fluorescens* CHA0, that acidic extracellular polysaccharides (EPS) are an important factor in adhesion to roots and fungi.

The ability to mark cells with different AFPs makes it possible to visualize simultaneously different species and populations. We have shown that it is possible to distinguish three different populations of *P. fluorescens* WCS365 cells tagged with *ecfp, egfp* and *rfp* simultaneously in the rhizosphere (Bloemberg et al. 2000). By looking at the ratios of differentially tagged cells present within the microcolonies it is suggested that most colonies are initiated by a single cell after which other cells from outside can still join the colony later in time as they will also leave colonies to colonize other parts of the root (Bloemberg et al. 2000). When visualizing two different organisms or populations a combination of GFP and DsRed (preferentially the improved forms) is most deficient since their spectra hardly overlap. Studies using mixed populations of two species, e.g., *P. chlororaphis* PCL1391 and *P. fluorescens* WCS365 showed that mixed colonies were formed, which were mostly present on the upper root part and that *P. chlororaphis* had a preference for colonizing the root hairs (Dekkers et al. 2000). The study of bacterial communities is also important to understand collaborations between bacteria as shown for instance for *P. putida* strains that are able to degrade polyaromatic hydrocarbons, which is stimulated by the presence of polyaromatic hydrocarbons suggesting that sensing processes, such as chemotaxis, are involved (Kuiper et al. 2002). Studies on phase variation of *Pseudomonas brassicacearum*, in which Phase I cells were labelled with GFP and Phase II cells with DsRed showed that Phase I and II cells colonize different niches on the root of *Arabidopsis thaliana* (Achouak et al. 2004). Whereas Phase I cells were mostly observed at the basal part of the root, Phase II cells were observed at secondary roots and root tips, which can be explained by the fact that Phase II cells make flagellin and are more motile. Also competition studies have been performed other than for *Pseudomonas* spp. such as for *S. meliloti* populations for which it was shown that mixed populations can be present in infection threads, which subsequently can result in mixed populations in the root nodule (Gage 2002).

Another valuable possibility for dual imaging is the visualization between biocontrol agents and phytopathogenic fungi that they control. After visualization of the infection process of *Fusarium oxysporum* f.sp. *radicis lycopersici* marked with GFP on tomato resulting in tomato foot and root rot (Lagopodi et al. 2002) interactions between *Pseudomonas* biocontrol species (tagged with *rfp*) and *Fusarium* were performed to obtain a better fundamental understanding of their interactions in the rhizosphere (Bolwerk et al. 2003) (Fig. 1, panels C and D). These studies showed that *Pseudomonas* and *Fusarium* compete for the colonization of same niches (intercellular junctions) and directly interact with each other. At sites where bacteria were present, infection of the root by penetration of *Fusarium* was not observed. In addition, *Pseudomonas* attached to the *Fusarium* hyphae and were able to extensively colonize those hyphae similar to what was observed during in vitro confrontation assays (Bolwerk et al. 2003). Molecular mechanisms involved in the attachment and colonization of the hyphae are hardly known. In the presence of *P. chlororaphis* PCL1391 many stress responses of the *Fusarium* were observed such as loss of growth directionality, increased vacuole formation, curly growth, swollen bodies and increased branching. It was shown that the antifungal compound produced by *P. chlororaphis* strain PCL1391 is (partially) responsible for these stress effects. We have also used GFP and CFP-expressing *Fusarium* strains to visualize the effect of biocontrol strains on *F. oxysporum*. Studies using *gfp* labelled *Trichoderma atroviride* and the (non-labelled) phytopathogenic fungi *Pythium ultimum* and *Rhizoctonia solani* on cucumber seeds showed that *T. atroviride* reacted to the presence of these pathogens by increased branching and the formation of morphological changes similar to hooks, appressoria, and papillae (Lu et al. 2004). AFPs will be an important tool for further elucidation of the molecular mechanisms involved in bacterium-fungus interactions.

Auto fluorescent proteins applied as tools for the visualization of ecological processes

Besides the purpose of localization of bacterial cells in their environment, auto fluorescent proteins are progressively used for other applications such as gene expression analysis, biosensor reporter systems, identification of specific environmental expressed genes and analysis of horizontal gene transfer in the environment.

The instable GFP variants made by Andersen et al. (1999) facilitate the analysis of transient gene expression in the rhizosphere. For instance, they were used to monitor the ribosomal activity of *P. putida* cells (Ramos et al. 2000). More recently, they were applied for the monitoring of anti-fungal metabolite production by the biocontrol agent *P. fluorescens* CHA0 (Baehler et al. 2005). Promoter regions of the biosynthetic genes of the three antifungal metabolites produced by CHA0, e.g., 2,4-diacetylphloroglucinol (DAPG), pyoluteorin (PLT) and pyrrolnitrin (PRN), were fused to GFP and their expression kinetics analysed in vitro. It will be exciting to apply these constructs in the rhizosphere and in the presence of fungi to analyze the influence of the fungus on the expression of the anti-fungal metabolite genes.

Stable GFP can also be used for expression studies in order to show that a gene of interest is expressed under certain conditions, without being able to analyze the transient expression profile. For example, it was shown by Rothballer et al. (2005) that the *ipdC* gene of *Azospirillum brasilense*, which is involved in indole-3-acetic acid production, is indeed expressed in the wheat rhizoplane. Other GFP-based expression systems were constructed to learn more about the direct interaction between the plant and the bacterium and the conditions that bacteria encounter in the rhizo- and phyllosphere. For example Leveau and Lindow (2001) applied an expression system that showed that the use of sugars, e.g., fructose and/or sucrose is mainly responsible for the growth of *Erwinia herbicola* on bean leaves. The construction and application of two bacterial sensors for the detection of nitrate availability indicated that roots compete for nitrate with the microbial rhizosphere population (DeAngelis et al. 2005). Studies on thiamine synthesis *gfp* reporter systems in *Rhizobium leguminosarum* bv. *viciae* showed that the rhizosphere of vetch and pea is poor in thiamine and that thiamine production is induced in the rhizosphere (Karunakaran et al. 2006). Another GFP-based study by Aldon et al. (2000) showed that physical contact between the bacterial cell and the plant strongly induces the expression of the *hrp* genes. The use of bioreporters has significantly contributed to the fundamental understanding of how bacteria interact with the plant (Leveau and Lindow 2002).

The development of GFP-based biosensor systems for the detection of AHL production were used to show that AHLs are produced in the microcolonies present in the rhizosphere, which will enable cross-talk between species in the rhizosphere (Andersen et al. 2001; Steidle et al. 2001).

A real challenge for researchers is to identify genes that are specifically expressed in the natural environment in order to understand the functioning of bacteria in their natural environment and to elucidate novel processes. In vivo expression technology is an elegant and powerful example of how environmentally-expressed genes can be identified (Rediers et al. 2005). In addition, other systems have been developed such as differential fluorescence induction (DFI) and optical trapping microscopy (Allaway et al. 2001), which for example allowed the identification of a rhizosphere-specific putative ABC transporter of putrescine (Allaway et al. 2001). Further development of systems for promoter trapping will facilitate the identification of other plant-associated genes (Izallalen et al. 2002).

When dual fluorescence imaging systems are not available, reporter systems based on two different techniques were shown to be of great value. For example, constructs carrying *gfp* and *gusA* in mini-Tn5 transposons (Xi et al. 1999) or plasmids (Ramos et al. 2002) were applied for the imaging of *Azospirillum brasilense* on wheat roots. These analyses showed that high concentrations of *Azospirillum* were present in the intercellular root cell spaces and at points of lateral root emergence, which are niches where nutrients presumably leak out of the root (Xi et al. 1999; Ramos et al. 2002). Other examples of combined reports are the use of immunofluorescence and a rRNA-targeting probe, which was used to analyze *P. fluorescens* DR54 in the sugar beet rhizosphere, showing that most metabolically active cells are found at the root tip and that endogenous soil microorganisms start colonizing the rhizosphere 2 days after seed inoculation (Lübeck et al. 2000). Another way to analyze metabolic activity is the use of luciferase, the activity of which is dependent on the cellular energy level. In combination with GFP, a construct containing *luxAB* genes was applied for *P. fluorescens* SBW25 showing that SBW25 was metabolically active in the whole rhizosphere (Unge et al. 1999; Unge and Jansson 2001), which can explain why the rhizosphere of wheat is more extensively colonized than, for example, the tomato rhizosphere.

As a last example I want to draw attention to recent work published by Molbak et al. (2006), in which horizontal gene transfer in the rhizosphere of pea and barley was analysed by CLSM studies. For that purpose they used *P. putida* KT2442 carrying a plasmid expressing *gfp*, which could be received by *P. putida* LM24 expressing *dsRed*. Their studies showed that plasmid transfer in the rhizosphere is determined by cell density and cell distribution, which are determined by exudation and root growth.

Future prospects

The application of auto fluorescent proteins will make a continuous contribution to the understanding of how PGPR function during their beneficial interactions with the plant and the endogenous microflora. Studying the molecular basis of the interactions between fungi and bacteria is an emerging field with great relevance in plant microbiology. Future studies will highly benefit from the developed tools for visualization of these organisms.

Acknowledgements I thank the microbiology group of the Institute of Biology Leiden (IBL) at Leiden University for fruitful discussions, their enthusiasm and results produced. I thank Gerda Lamers of the IBL microscopy unit for her valuable advise on the optimal use of confocal laser scanning microscopy and her technical assistance.

References

Achouak, W., Conrod, S., Cohen, V., & Heulin, T. (2004). Phenotypic variation of *Pseudomonas brassicacearum* as a plant root-colonization strategy. *Molecular Plant Microbe Interactions, 17*, 872–879.

Aldon, D., Brito, B., Boucher, C., & Genin, S. (2000). A bacterial sensor of plant cell contact controls the transcriptional induction of *Ralstonia solanacearum* pathogenicity genes. *EMBO Journal, 19*, 2304–2314.

Allaway, D., Schofield, N. A., Leonard, M. E., Gilardoni, L., Finan, T. M., & Poole, P. S. (2001). Use of differential fluorescence induction and optical trapping to isolate environmentally induced genes. *Environmental Microbiology, 3*, 397–406.

Andersen, J. B., Heydorn, A., Hentzer, M., Eberl, L., Geisenberg, O., Christensen, B. B., Molin, S., & Givskov, M. (2001). *Gfp*-Based *N*-acyl homoserine-lactone sensor systems for detection of bacterial communities. *Applied and Environmental Microbiology, 67*, 575–585.

Andersen, J. B., Sternberg, C., Poulsen, L. K., Bjorn, S. P., Givskov, M., & Molin, S. (1999). New unstable variants of green fluorescent protein for studies of transient gene expression in bacteria. *Applied and Environmental Microbiology, 64*, 2240–2246.

Baehler, E., Bottiglieri, M., Pechy-Tarr, M., Maurhofer, M., & Keel, C. (2005). Use of green fluorescent protein-based reporters to monitor balanced production of antifungal compounds in the biocontrol agent *Pseudomonas fluorescens* CHA0. *Journal of Applied Microbiology, 99*, 24–38.

Biancotto, V., Andreotti, S., Balestrini, R., Bonfante, P., & Perotto, S. (2001). Extracellular polysaccharides are involved in the attachment of *Azospirillum brasilense* and *Rhizobium leguminosarum* to arbuscular mycorrhizal structures. *European Journal of Histochemistry, 45*, 39–49.

Bloemberg, G. V., & Camacho, M. (2006). Microbial interactions with plants; a hidden world? In: B. Schultz, C. Boyle & T. Sieber (Eds.), *Soil biology, volume 9, microbial root endophytes* (pp. 321–336). Heidelberg: Springer Verlag Berlin.

Bloemberg, G. V., Lagopodi A., de Bruijn, F. J., & Jansson, J. K. (2004). Visualisation of microbes and their interactions in the rhizosphere using auto fluorescent proteins as markers. In: G. A. Kowalchuk, F. J. de Bruijn, I. M Head, A. D. Akkermans & J. D. v. Elsas (Eds.), *Molecular microbial ecology manual* (pp. 1257–1280). Heidelberg: Springer Berlin.

Bloemberg, G. V., & Lugtenberg, B. J. J. (2004). Biofilm formation on plants, their relevance and phenotypic aspects. In: Ghannoum, & G. O'Toole (Eds.). *Microbial biofilms* (pp. 141–159). Washington, D.C.: ASM press.

Bloemberg, G. V, Wijfjes, A. H. M., Lamers, G. E. M., Stuurman, N., & Lugtenberg, B. J. J. (2000). Simultaneous imaging of *Pseudomonas fluorescens* WCS365 populations expressing three different autofluorescent proteins in the rhizosphere; new perspectives for studying microbial communities. *Molecular Plant Microbe Interactions, 13*, 1170–1176.

Bloemberg, G. V., O'Toole, G., Lugtenberg, B. J. J., & Kolter, R. (1997). Green fluorescent protein as a marker for *Pseudomonas* spp. *Applied and Environmental Microbiology, 63*, 4543–4551.

Bolwerk, A., Lagopodi, A. L., Wijfjes, A. H. M., Lamers, G. E. M., Chin-A-Woeng, T. F. C., Lugtenberg, B. J. J., & Bloemberg, G. V. (2003). Interactions in the tomato rhizosphere of two *Pseudomonas* biocontrol strains with the phytopathogenic fungus *Fusarium oxysporum* f. sp. *radicis-lycopersici*. *Molecular Plant-Microbe Interactions, 16*, 983–993.

Burlage, R. S., Yang, Z. K., & Mehlhorn, T. (1995). A transposon for green fluorescent protein transcriptional fusions: Applications for bacterial experiments. *Gene, 173*, 53–58.

Chalfie M., Kain S. R. (Eds.) (2005). *Green fluorescent protein: Properties, applications and protocols* (*Methods of Biochemical Analysis, Vol 47*). New York: Wiley.

Chalfie, M., Tu, Y., Euskirchen, G., Ward, W. W., & Prasher, D. C. (1994). Green fluorescent protein as a marker for gene expression. *Science, 263*, 802–805.

Dandie, C. E., Larrainzar, E., Mark, G. L., O'Gara, F., & Morrisey, J. P. (2005). Establishment of DsRed.T3_S4T as an improved autofluorescent marker for microbial ecology applications. *Environmental Microbiology, 7*, 1818–1825.

DeAngelis, K. M., Ji, P., Firestone, M. K., & Lindow, S. E. (2005). Two novel bacterial biosensors for detection of nitrate availability in the rhizosphere. *Applied and Environmental Microbiology, 71*, 8537–8547.

de Groot, M. J., Bundock, P., Hooykaas, P. J., & Beijersbergen, A. G. (1998). *Agrobacterium tumefaciens*-mediated transformation of filamentous fungi. *Nature Biotechnology, 16*, 839–842.

Dekkers, L. C., Mulders, I. H., Phoelich, C. C., Chin-A-Woeng, T. F. C., Wijfjes, A. H. M., & Lugtenberg, B. J. J. (2000). The *sss* colonization gene of the tomato-*Fusarium oxysporum* f.sp. *radicis lycopersici* biocontrol strain *Pseudomonas fluorescens* WCS365 can improve root colonization of other wild-type *Pseudomonas* spp. bacteria. *Molecular Plant Microbe Interactions, 13*, 1177–1183.

Elbeltagy, A., Nishioka, K., Sato, T., Suzuki, H., Ye, B., Hamada, T., Isawa, T., Mitsui, H., & Minamisawa, K. (2001). Endophytic colonization and in planta nitrogen fixation by a *Herbaspirillum* sp. isolated from wild rice species. *Applied and Environmental Microbiology, 67*, 5285–5293.

Ellenberg, J., Lippincott, S. J., & Presley, J. F. (1999). Dual-colour imaging with GFP variants. *Trends in Cell Biology, 9*, 52–56.

Gage, D. J. (2002). Analysis of infection thread development using *Gfp*- and DsRed-expressing *Sinorhizobium meliloti*. *Journal of Bacteriology, 184*, 7042–7046.

Gage, D. J., Bobo, T., & Long, S. R. (1996). Use of green fluorescent protein to visualize the early events of symbiosis between *Rhizobium meliloti* and alfalfa (*Medicago sativa*). *Journal of Bacteriology, 178*, 7159–7166.

Gamalero, E., Lingua, G., Tombolini, R., Avidano, L., Pivato, B., & Berta, G. (2005). Colonization of tomato root seedling by *Pseudomonas fluorescens* 92 rkG5: Spatiotemporal dynamics, localization, organization, viability, and culturability. *Microbial Ecology, 50*, 289–297.

Gotz, M., Gomes, N. C., Dratwinski, A., Costa, R., Berg, G., Peixoto, R., Mendonca-Hagler, L., & Smalla, K. (2006). Survival of *gfp*-tagged antagonistic bacteria in the rhizosphere of tomato plants and their effects on the indigenous bacterial community. *FEMS Microbiology Ecology, 56*, 207–218.

Heeb, S., Itoh, Y., Nishijyo, T., Schnider, U., Keel, C., Wade, J., Walsh, U., O'Gara, F., & Haas, D. (2000). Small, stable shuttle vectors based on the minimal pVS1 replicon for use in gram-negative, plant-associated bacteria. *Molecular Plant-Microbe Interactions, 13*, 232–237.

Heim, R., Prasher, D. C., & Tsien, R. Y. (1994). Wavelength mutations and post-translational autoxidation of green fluorescent protein. *Proceedings National Academy of Sciences USA., 91*, 12501–12504.

Izallalen M., Levesque, R. C., Perret, X.., Broughton, W. J., & Antoun, H. (2002). Broad-host-range mobilizable suicide vectors for promoter trapping in gram-negative bacteria. *Biotechniques, 33*, 1038–1043.

Karunakaran, R., Ebert, K., Harvey, S., Leonard, M. E., Ramachandran, V., & Poole, S. P. (2006). Thiamine is synthesized by a salvage pathway in *Rhizobium leguminosarum* bv. *viciae* strain 3841. *Journal of Bacteriology, 188*, 6661–6668.

Kuiper, I., Kravchenkov, L., Bloemberg, G. V., & Lugtenberg, B. J. J. (2002). *Pseudomonas putida* strain PCL1444, selected for efficient root colonization and naphthalene degradation, effectively utilizes root exudates components. *Molecular Plant Microbe Interactions, 15*, 734–741.

Lagopodi, A. L., Ram, A. F. J., Lamers, G. E. M., Punt, P. J., van den Hondel, C. A. M. J., Lugtenberg, B. J. J., & Bloemberg, G. V. (2002). Confocal laser scanning microscopical analysis of tomato root colonization and infection by *Fusarium oxysporum* f.sp. *radicis-lycopersici* using the green fluorescent protein as a marker. *Molecular Plant Microbe Interactions, 15*, 172–179.

Larrainzar, E., O'Gara, F., & Morrisey, J. P. (2005). Applications of autofluorescent proteins for in situ studies in microbial ecology. *Annual Review of Microbiology, 59*, 257–277.

Leveau, J. H., & Lindow, S. E. (2001). Appetite of an epiphyte: Quantitative monitoring of bacterial sugar consumption in the phyllosphere. *Proceedings of the National Academy of Sciences USA, 98*, 3446–3453.

Leveau, J. H., & Lindow, S. E. (2002). Bioreporters in microbial ecology. *Current Opinion in Microbiology, 5*, 259–265.

Lübeck, P. S., Hansen, M., & Sørensen, J. (2000). Simultaneous detection of the establishment of seed-inoculated *Pseudomonas fluorescens* strain Dr54 and native soil bacteria on sugar beet root surfaces using fluorescense antibody and in situ hybridization techniques. *FEMS Microbiology Ecology, 33*, 11–19.

Lu, Z., Tombolini, R., Woo, S., Zeilinger, S., Lorito, M., & Jansson, J. K. (2004). In vivo study of trichoderma-pathogen-plant interactions, using constitutive and inducible green fluorescent protein reporter systems. *Applied and Environmental Microbiology, 70*, 3073–3081.

Matus, A. (1999). GFP in motion CD-ROM - Introduction: GFP illuminates everything. *Trends in Cell Biology, 9*, 43.

Matz, M. M. V., Fradkov, A. F., Labas, Y.A., Savitsky, A. P., Zaraisky, A. G., Markelov, M. L., & Lukyanov, S. A. (1999). Fluorescent proteins from non-bioluminescent *Anthozoa* species. *Nature Biotechnology, 17*, 969–973.

Michielse, C. B., Hooykaas, P. J., van den Hondel, C. A., & Ram A. F. (2005). *Agrobacterium*-mediated transformation as a tool for functional genomics in fungi. *Current Genetics, 48*, 1–17.

Molbak, L., Molin, S., & Kroer, N. (2006). Root growth and exudate production define the frequency of horizontal plasmid transfer in the rhizosphere. *FEMS Microbiology Ecology, 59*, 167–176.

Newman, K. L., Almeida, R. P., Purcell, A. H., & Lindow, S. E. (2003). Use of a green fluorescent strain for analysis of *Xylella fastidiosa* colonization of *Vitis vinifera*. *Applied and Environmental Microbiology, 69*, 7319–7327.

Ramos, H. J., Roncato-Maccari, L. D., Souza, F. M., Soares-Ramos, J. R., Hungria, M., & Pedrosa, F. O. (2002). Monitoring *Azospirillum*-wheat interactions using the *gfp* and *gusA* genes constitutively expressed from a new broad-host range vector. *Journal of Biotechnology, 97*, 243–252.

Ramos, C., Molbak, L., & Molin, S. (2000). Bacterial activity in the rhizosphere analyzed at the single-cell level by monitoring ribosome contents and synthesis rates. *Applied and Environmental Microbiology, 66*, 801–809.

Rediers, H., Rainey, P. B., Vanderleyden, J., & De Mot, R. (2005). Unraveling the secret lives of bacteria: Use of in vivo expression technology and differential fluorescence induction promoter traps as tools for exploring niche specific gene expression. *Microbiology and Molecular Biology Reviews, 69*, 217–261.

Rothballer, M., Schmid, M., Fekete, A., & Hartmann, A. (2005). Comparative in situ analysis of *ipdC-gfpmut3* promoter fusions of *Azospirillum brasilense* strains Sp7 and Sp245. *Environmental Microbiology, 7*, 1839–1846.

Shaner, N. C., Campbell, R. E., Steinbach, P. A., Giepmans, B. N. G., Palmer, A. E., & Tsien, R. Y. (2004). Improved monomeric red, orange and yellow fluorescent proteins derived from *Discosoma* sp. red fluorescent protein. *Nature Biotechnology, 22*, 1567–1572.

Sörensen, M., Lippuner, C., Kaiser, T., Misslitz, A., Aebischer, T., & Bumann, D. (2003). Rapidly maturing red fluorescent protein variants with strongly enhanced brightness in bacteria. *FEBS Letters, 552*, 110–114.

Steidle, A., Sigl, K., Schuhegger, R., Ihring, A., Schmid, M., Gantner, S., Stoffels, M., Riedel, K., Givskov, M., Hartman, A., Langebartels, C., & Eberl, L. (2001). Visualisation of *N*-acylhomoserine lactone-mediated cell-cell communication between bacteria colonizing the tomato rhizosphere. *Applied and Environmental Microbiology, 67*, 5761–5770.

Stuurman, N., Pacios Bras, C., Schlaman, C. H. R. M., Wijfjes, A. H. M., Bloemberg, G. V., & Spaink, H. P. (2000). The use of GFP color variants expressed on stable broad-host range vectors to visualize rhizobia interacting with plants. *Molecular Plant-Microbe Interactions, 13*, 1063–1069.

Tombolini, R., Unge, A., Davey, M. E., de Bruijn, F. J., & Jansson, J. K. (1997). Flow cytometric and microscopic analysis of GFP-tagged *Pseudomonas fluorescens* bacteria. *FEMS Microbiology Ecology, 22*, 17–28.

Tombolini, R., van der Gaag, D. J., Gerhardson, B., & Jansson, J. K. (1999). Colonization pattern of the biocontrol strain *Pseudomonas chlororaphis* MA342 on barley seeds visualized by using green fluorescent protein. *Applied and Environmental Microbiology, 65*, 3674–3680.

Tsien, R. Y. (1998). The green fluorescent protein. *Annual Reviews of Biochemistry, 67*, 509–544.

Unge, A., Tombolini, R., Moller, A., & Jansson, J. K. (1997). Optimization of GFP as a marker for detection of bacteria in environmental samples. In: J. W. Hastings, L. J. Kricka & P. E. Stanley (Eds.). 9th International Symposium on bioluminescence and chemiluminescence: Proceedings volume: Bioluminescence and Chemiluminescence: Molecular Reporting With Photons.

Unge, A., Tombolini, R., Molbak, L., & Jansson, J. K. (1999). Simultaneous monitoring of cell number and metabolic activity of specific bacterial populations with a dual *gfp-luxAB* marker system. *Applied and Environmental Microbiology, 65*, 813–821.

Unge, A., & Jansson, J. K. (2001). Monitoring population size, activity, and distribution of *gfp-luxAB*-tagged *Pseudomonas fluorescens* SBW25 during colonization of wheat. *Microbiology Ecology, 41*, 290–300.

Utermark, J., & Karlovsky, P. (2006). Quantification of green fluorescent protein fluorescence using real-time PCR thermal cycler. *Biotechniques, 41*, 150–154.

Xi, C., Lambrecht, M., Vanderleyden, J., & Michiels, J. (1999). Bi-functional *gfp*- and *gusA*-containing mini-Tn5 transposon derivatives for combined gene expression and bacterial localization studies. *Journal of Microbiological Methods, 35*, 85–92.

Yang, T. T., Sina, P., Green, G., Kitts, P. A., Chen, Y. T., Lybarger, L., Chervenak, R., Patterson, G. H., Piston, D. W., & Kain, S. R. (1998). Improved fluorescence and dual color detection with enhanced blue and green variants of the green fluorescent protein. *Journal of Biological Chemistry, 273*, 8212–8216.

Eur J Plant Pathol (2007) 119:311–328
DOI 10.1007/s10658-007-9157-1

REVIEW PAPER

Dialogues of root-colonizing biocontrol pseudomonads

Christophe Dubuis · Christoph Keel · Dieter Haas

Received: 21 December 2006 / Accepted: 26 April 2007 / Published online: 5 June 2007

Abstract Among biocontrol agents that are able to suppress root diseases caused by fungal pathogens, root-colonizing fluorescent pseudomonads have received particular attention because many strains of these bacteria trigger systemic resistance in host plants and produce antifungal compounds and exoenzymes. In general, the expression of these plant-beneficial traits is regulated by autoinduction mechanisms and may occur on roots when the pseudomonads form microcolonies. Three major classes of antibiotic compounds reviewed here in detail (2,4-diacetylphloroglucinol, pyoluteorin and various phenazine compounds) are all produced under cell population density-dependent autoinduction control acting at transcriptional and post-transcriptional levels. This regulation can either be reinforced or attenuated by a variety of chemical signals emanating from the pseudomonads themselves, other microorganisms or root exudates. Signals stimulating biocontrol factor expression via the Gac/Rsm signal transduction pathway in the biocontrol strain *Pseudomonas fluorescens* CHA0 are synthesized by many different plant-associated bacteria, warranting a more detailed investigation in the future.

C. Dubuis · C. Keel · D. Haas (✉)
Département de Microbiologie Fondamentale,
Université de Lausanne, Lausanne 1015, Switzerland
e-mail: Dieter.Haas@unil.ch

Keywords Antibiotic compounds · Biocontrol · GacS/GacA · Quorum sensing · *Pseudomonas* · Rhizosphere

Abbreviations

AHL	*N*-acyl-homoserine lactone
AI-2	Autoinducer 2
DAPG	2,4-Diacetylphloroglucinol
HCN	Hydrogen cyanide
IAA	Indole-3-acetic acid
PCA	Phenazine-1-carboxylic acid
PLT	Pyoluteorin
PQS	*Pseudomonas* quinolone signal
QS	Quorum sensing

Introduction

Probably all free-living bacteria communicate with one another by synthesizing, secreting and sensing signal molecules, which diffuse in the environment and are detected by specific cellular receptors. The main tasks of these signals are to coordinate metabolic activities and developmental processes in bacterial populations. In general, the higher the cell population densities are in growing cells, the more important signalling becomes. This relationship is known as quorum sensing (QS) and it is assumed that signalling enables the producer bacteria to assess their population densities (Fuqua et al. 1994). To some extent, this is certainly an important aspect. In many cases, signals amplify their own biosynthesis. Thus, in growing populations, this positive feedback loop results in autoinduction and for this reason, such QS signals are also termed autoinducers (Nealson 1977). However, the physiological condition of the cells and many environmental factors can modulate the concentration of the QS signals and hence the cellular response (Lazdunski et al. 2004; Hense et al. 2007).

In the rhizosphere, bacterial populations preferentially form microcolonies in areas that are particularly rich in root exudates (Bais et al. 2006). Here, the production of QS signals can readily be detected by signal-specific biosensors (Pierson et al. 1998; Steidle et al. 2001). The calling distance, i.e. the range of cell-cell communication via chemical signals, has been experimentally determined on plant roots and estimated to be of the order of 5 μm, but under favourable conditions, e.g. at root tips or root hair junctions, it can extend beyond 50 μm (Gantner et al. 2006). In many rhizobacteria, QS mechanisms induce the synthesis of antimicrobial secondary metabolites and extracellular lytic enzymes that are inhibitory to other bacteria, fungi, protozoa, and nematodes (Haas et al. 2000; Haas and Défago 2005; Siddiqui et al. 2005; Juhas et al. 2005; Jousset et al. 2006). We have argued elsewhere that a primary function of these antimicrobial factors is to protect the producer microcolonies from invasion by other inhabitants of the rhizosphere and hence to preserve the integrity of the producer in its ecological niche on plant roots (Haas and Keel 2003). The measured calling distances of QS signals correspond well to microcolony dimensions. When antibiotic-producing rhizobacteria antagonize the proliferation of root-pathogenic fungi on roots and when, in addition, the defence reactions of the host plant are stimulated, a significant biocontrol effect can result: the plant becomes less susceptible to disease caused by pathogenic fungi. In this review, we will discuss some signal transduction pathways that positively control the expression of biocontrol factors in root-colonizing soil bacteria belonging to the group of fluorescent pseudomonads. These bacteria have been studied in considerable detail for their biocontrol properties and their signalling pathways (Moënne-Loccoz and Défago 2004; Juhas et al. 2005; Haas and Défago 2005).

Signals and nutrients

Signals that are engaged in cell-cell communication of microorganisms meet several criteria: they are produced during a particular growth phase and in response to environmental conditions, they accumulate in the extracellular medium, they interact specifically with cellular receptors and, above a threshold, they trigger expression of a set of genes. However, all these criteria can also be fulfilled by nutrients which are degraded by microorganisms through inducible pathways. For instance, glucose, a sugar present in the rhizosphere (Lugtenberg and Bloemberg 2004), is converted to gluconate by glucose oxidase in fluorescent pseudomonads (Quay et al. 1972). The diffusible product gluconate, which acts as an antifungal agent (Kaur et al. 2006), induces the enzymes of the Entner-Doudoroff pathway leading to the degradation of gluconate and glucose (Quay et al. 1972). Should gluconate be considered a signal? Some authors indeed do not discriminate between rhizosphere signals that primarily serve as nutrients and those that specifically trigger gene expression without having nutritional value (Somers et al. 2004). Other authors emphasize that signals used in cell-cell communication typically regulate gene expression beyond the physiological changes required to metabolize or to inactivate the signals and that, therefore, signals should be distinguished from nutrients (Winzer et al. 2002). In the following, we will adopt the latter view and we will not discuss how nutrients present in exudates chemotactically attract soil bacteria and how these bacteria subsequently attach to root surfaces and form microcolonies. These processes have been reviewed elsewhere (Somers et al. 2004; Lugtenberg and Bloemberg 2004).

Signals that influence the expression of biocontrol traits in pseudomonads can emanate from the biocontrol bacteria themselves, from other soil bacteria or fungi, or from host plants (Fig. 1). Like any biological molecules, signals can be degraded, either by the producing organisms, which thereby avoid an overshooting response (Huang et al. 2006a; Wang et al. 2006), or by other organisms which cause interference of signalling pathways (Zhang and Dong 2004; Xavier and Bassler 2005; González and Keshavan 2006). Signals can have stimulating or inhibitory effects, with repercussion on transcriptional or translational control of biocontrol gene expression.

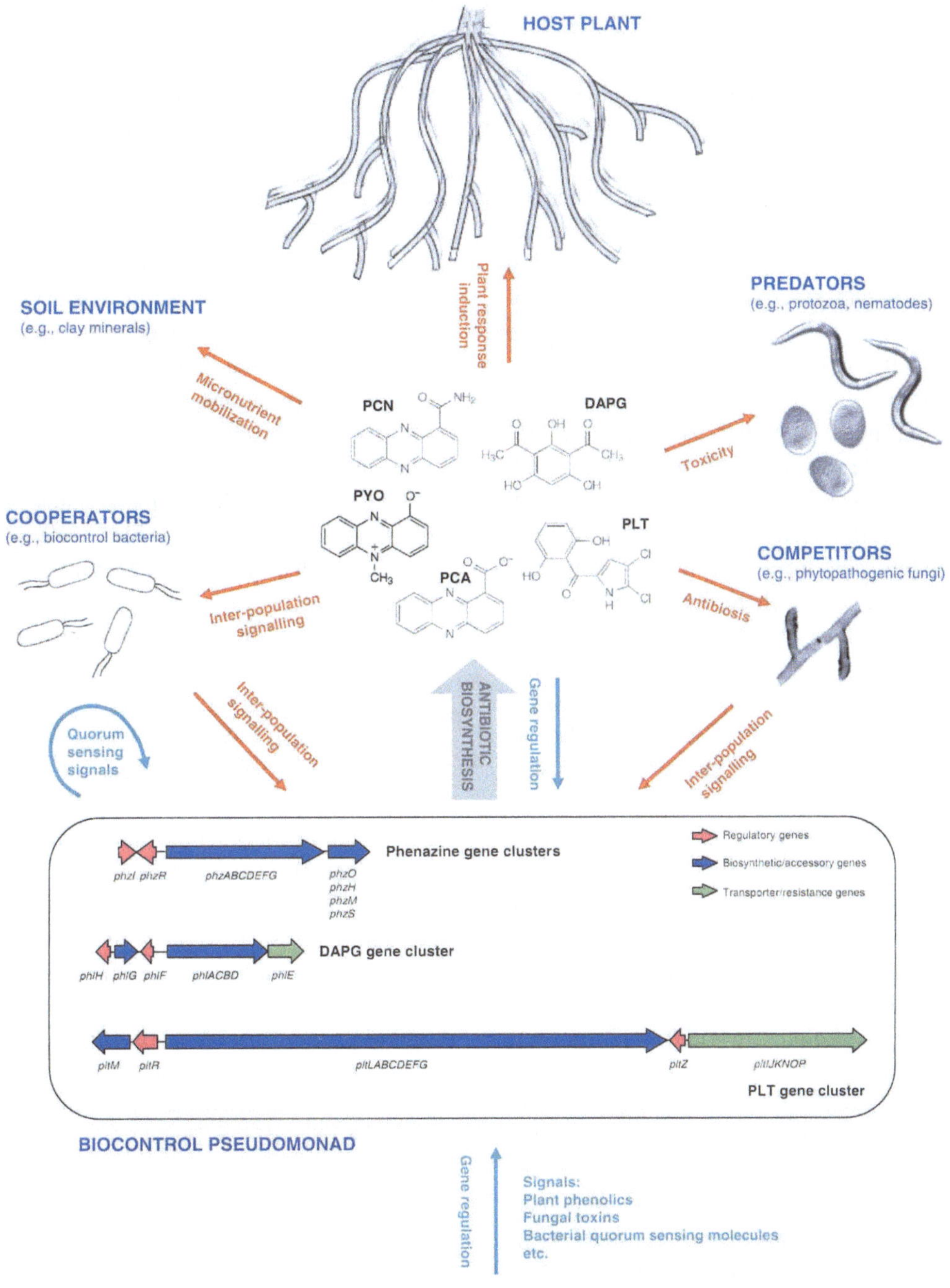

Fig. 1 Overview of interactions between biocontrol strains, plants, pathogens, predators, co-operators, and soil. The antibiotics (phenazines, DAPG, PLT) produced by the biocontrol strains play a central role between all the elements of the complex network of interactions. The organization of the *phz*, *phl* and *plt* biosynthetic genes in fluorescent pseudomonads is according to Mavrodi et al. (2006); Bangera and Thomashow (1999); Schnider-Keel et al. (2000); and Brodhagen et al. (2005). Red arrows designate interactions, green arrows designate transcriptional gene regulation

Signals affecting biocontrol factor expression at a transcriptional level

Control of root diseases by biocontrol pseudomonads involves a blend of complementary mechanisms, the most prominent being antibiosis towards plant pathogens, degradation of virulence factors produced by pathogens and induction of defence mechanisms in host plants. Efficient competition for colonization sites and micro- and macro-nutrients in the rhizosphere is an important prerequisite for effective biocontrol (van Loon et al. 1998; Lugtenberg et al. 2001; Compant et al. 2005; Haas and Défago 2005). Root-colonizing plant-beneficial pseudomonads release a remarkable diversity of metabolites with antibiotic activity. There is compelling experimental evidence from genetic analyses involving antibiotic-negative mutants and from detection of antibiotic production in the rhizosphere by analytical and reporter gene techniques that these compounds are important for biocontrol activity (Raaijmakers et al. 2002; Haas and Keel 2003; Haas and Défago 2005). Furthermore, it has been established that populations of antibiotic-producing, rhizosphere-associated pseudomonads are key biological components in natural soils that are suppressive to major root diseases such as take-all of wheat and black root of tobacco (Keel et al. 1996; Weller et al. 2002; Ramette et al. 2003; de Souza et al. 2003a; Weller et al. 2007).

Among the antimicrobial compounds released by plant-beneficial pseudomonads, 2,4-diacetylphloroglucinol (DAPG), pyoluteorin (PLT), phenazines, hydrogen cyanide (HCN), and cyclic lipopeptides (e.g., viscosinamide, amphisin, putisolvin) have received particular attention for their major contribution to biocontrol of root diseases that are caused by agronomically important fungal and oomycete pathogens including *Gaeumannomyces*, *Thielaviopsis*, *Fusarium*, and *Pythium* species (Haas and Keel 2003; Chin-A-Woeng et al. 2003; Andersen et al. 2003; Kuiper et al. 2004; Mavrodi et al. 2006; Raaijmakers et al. 2006; Rezzonico et al. 2007; Weller et al. 2007). Most effective biocontrol pseudomonads produce at least one of the above-mentioned diffusible or volatile antibiotics and some strains, e.g. *Pseudomonas fluorescens* strains CHA0 and Pf-5, produce multiple antibiotics. In general, these antibiotics have broad-spectrum toxic activity against fungi, bacteria, protozoa, nematodes, and sometimes also against plants or even viruses (Raaijmakers et al. 2002, 2006; Haas and Keel 2003; Jousset et al. 2006). However, for DAPG and PLT the precise modes of action are largely unknown. Phenazines being redox-active antibiotics are thought to harm phytopathogenic fungi by generation of reactive oxygen species (Chin-A-Woeng et al. 2003; Mavrodi et al. 2006; Price-Whelan et al. 2006). HCN is an inhibitor of metalloenzymes, in particular of most terminal oxidases (Blumer and Haas 2000). Cyclic lipopeptides are surfactants promoting surface motility of the producer organism; at high concentrations, they are detrimental to the integrity of phospholipid membranes in various organisms (Raaijmakers et al. 2006).

Some of the antibiotic metabolites have remarkably diverse functions, besides their toxic activity (Fig. 1). DAPG, PLT, and phenazines can function as signal molecules that affect gene expression not only in the producer bacteria, but also in other organisms (Schnider-Keel et al. 2000; Maurhofer et al. 2004; Brodhagen et al. 2004; Baehler et al. 2005; Dietrich et al. 2006; Price-Whelan et al. 2006). DAPG has been described as an inducer of systemic plant resistance (Iavicoli et al. 2003; Weller et al. 2007) and as a stimulant of amino acid exudation from roots (Phillips et al. 2004). Phenazines, in their reduced form, might enable the producing bacteria to mobilize micronutrients such as iron (Fe^{3+}) from the rhizosphere environment (Hernandez et al. 2004; Price-Whelan et al. 2006). In the following, we will highlight some examples of signal perception and exchange involving transcriptional control mechanisms, with a focus on the biosyntheses of DAPG, PLT, and phenazines (Fig. 1).

DAPG and PLT as signals

DAPG has been shown to be particularly important in the suppression of root diseases caused by *Thielaviopsis basicola* (Keel et al. 1990), *Pythium ultimum* (Fenton et al. 1992; Baehler et al. 2006) and *Gaeumannomyces graminis* (Keel et al. 1992). The DAPG biosynthetic locus has been identified and analysed in some detail in *P. fluorescens* strains Q2-87, CHA0, and F113 (Bangera and Thomashow 1999; Schnider-Keel et al. 2000; Delany et al. 2000). The DAPG locus comprises the biosynthetic genes *phlACBD*; *phlD* encodes a type III polyketide synthase catalysing the synthesis of phloroglucinol from malonyl-CoA and *phlACB* encodes enzymes that presumably are required for the subsequent acetylation reactions to mono-acetylphloroglucinol and DAPG (Achkar et al. 2005; Zha et al. 2006). The *phlE* gene located downstream of the *phlACBD* cluster codes for a putative transport/resistance protein (Bangera and Thomashow 1999; Abbas et al. 2004). The *phlF, phlG*, and *phlH* genes located upstream of the biosynthetic genes encode, respectively, a pathway-specific transcriptional repressor belonging to the TetR family, a hydrolase that specifically degrades DAPG to less toxic mono-acetylphloroglucinol, and a second TetR-type transcriptional regulator that activates the expression of the *phlACBD* genes and negatively controls *phlG* (Bangera and Thomashow 1999; Schnider-Keel et al. 2000; Delany et al. 2000; Abbas et al. 2004; Bottiglieri and Keel 2006). Pathway-specific control in strains CHA0 and F113 is brought about by the PhlF protein which represses the expression of the *phlACBD* operon by binding to two conserved operator sites in the *phlA* leader region (Schnider-Keel et al. 2000; Delany et al. 2000; Haas and Keel 2003; Abbas et al. 2004; D. Haas and C. Gigot-Bonnefoy, unpublished results). DAPG itself acts as the derepressing signal by dissociating the repressor PhlF from the *phlA* promoter, thereby acting as an autoinducer of its own biosynthesis (Schnider-Keel et al. 2000; Abbas et al. 2004). PhlH, the second pathway-associated transcriptional regulator, is postulated to antagonize the repressive effect of PhlF, but the precise mechanism remains to be determined (Haas and Keel 2003).

Pyoluteorin is another important antimicrobial metabolite that is produced by some *Pseudomonas* strains and that contributes in particular to the suppression of root diseases caused by the oomycete *P. ultimum* (Howell and Stipanovic 1980; Maurhofer et al. 1992, 1994). The PLT biosynthetic locus has been identified in *P. fluorescens* Pf-5 (Kraus and Loper 1995; Nowak-Thompson et al. 1999), in *P. fluorescens* CHA0 (Péchy-Tarr et al. 2005) and in *Pseudomonas* sp. M18, which appears to belong to *P. aeruginosa* (Ge et al. 2004). The biosynthetic locus comprises the *pltLABCDEFG* and *pltM* genes; *pltB* and *pltC* encode polyketide synthases with core functions in PLT biosynthesis (Nowak-Thompson et al. 1999). Pathway-specific positive control of PLT biosynthetic gene expression in strains CHA0 and Pf-5 is brought about by a LysR-type transcriptional regulator encoded by the divergently transcribed *pltR* gene (Nowak-Thompson et al. 1999; M. Bottiglieri and C. Keel, unpublished data). PLT induces its own biosynthesis via an autoinduction mechanism similar to that described for DAPG (Brodhagen et al. 2004; Baehler et al. 2005) and there is some experimental evidence that the transcriptional activator PltR may mediate this mechanism (M. Bottiglieri and C. Keel, unpublished data). Recently, a second regulatory gene, *pltZ*, linked to the PLT biosynthetic locus has been identified in strain M18 (Huang et al. 2004) and *P. fluorescens* strain Pf-5 (Brodhagen et al. 2005). This gene encodes a TetR-type transcriptional repressor of PLT gene expression (Huang et al. 2004) and is located upstream of a genomic region (*pltHIJKNO* in strain M18 and *pltIJKNOP* in strain Pf-5) encoding an ABC-type transport apparatus for PLT and accessory membrane-bound proteins (Brodhagen et al. 2005; Huang et al. 2006b). Exogenous PLT induces the expression of both the PLT biosynthetic and transport operons, suggesting that PLT production and export are coordinated (Brodhagen et al. 2005; Huang et al. 2006a, b).

In *P. fluorescens* strains CHA0 and Pf-5, a mechanism of molecular cross-talk between the DAPG and PLT biosynthetic pathways has been identified that enables the bacteria to maintain production of these key antibiotics at balanced levels. In this molecular balance, DAPG and PLT act as signalling molecules inducing the expression of their own biosynthetic genes while strongly repressing the expression of the biosynthetic genes of the other antibiotic (Schnider-Keel et al. 2000; Brodhagen et al. 2004; Baehler et al. 2005, 2006). The pathway-specific transcriptional regulators PhlF and PltR appear to function as receptors of these signals. However, the underlying molecular mechanisms still need to be substantiated. The housekeeping sigma factor RpoD, the alternative sigma factors RpoS, RpoN and RpoE (Schnider et al. 1995; Sarniguet et al. 1995; Johansen et al. 2002; Haas and Keel 2003; Péchy-Tarr et al. 2005), Lon protease (Whistler et al. 2000), and the H-NS-like proteins MvaT and MvaV (Baehler et al. 2006) influence the DAPG/PLT balance. In the rhizosphere environment, these global transcriptional regulators might allow the bacteria to coordinate antibiotic production with other cellular functions, such as nutrition acquisition and response to environmental stress.

As exogenous signals, DAPG and PLT mediate both intra- and interpopulation communication. In the genetically distinct biocontrol strains CHA0 and Q2-87, DAPG produced by either strain on wheat roots is perceived as a positive signal boosting in situ expression of *phl* genes in the other strain (Maurhofer et al. 2004). Similarly, PLT produced by *P. fluorescens* Pf-5 acts as a signal activating *plt* gene expression in a PLT-negative derivative of Pf-5 in the rhizosphere of cucumber (Brodhagen et al. 2004). Salicylate, a metabolite produced by strain CHA0, numerous other pseudomonads and plants, strongly represses the production of DAPG and PLT and the expression levels of the corresponding biosynthetic genes (Schnider-Keel et al. 2000; Baehler et al. 2005). It is noteworthy that QS molecules of the *N*-acyl-homoserine lactone (AHL) type, which are synthesized by some DAPG producers such as *P. fluorescens* F113 (Laue et al. 2000) and 2P24 (Wei and Zhang 2006), do not contribute to the regulation of antibiotic biosynthesis in these bacteria, although AHLs function as key signals in cell-density dependent regulation of phenazine antibiotic production in other biocontrol pseudomonads (see below).

Whereas pathogenic fungi are major targets of biocontrol pseudomonads, only few examples of specific signalling interactions between these organisms have been documented. Fusaric acid, a toxin and pathogenicity factor of the root pathogen *Fusarium oxysporum*, is probably the most prominent example of a fungal compound that interferes with bacterial expression of a biocontrol trait. Fusaric acid strongly represses DAPG production in *P. fluorescens* CHA0 (Duffy and Défago 1997; Schnider-Keel et al. 2000; Notz et al. 2002). This repression may ultimately result in failure of the bacterium to suppress the root disease caused by the fungus (Duffy and Défago 1997; Duffy et al. 2004). An example of positive signalling is the stimulation of DAPG gene expression in strain CHA0 by unidentified signals from culture filtrates and volatiles of the biocontrol fungus *Trichoderma atroviride* (Lutz et al. 2004). Several fungal pathogens including *Gaeumannomyces*, *Pythium*, and *Fusarium* cope with DAPG by detoxifying it or by developing other mechanisms of resistance to it (Mazzola et al. 1995; de Souza et al. 2003b; Duffy et al. 2003; Schouten et al. 2004).

Little is known about plant signals affecting the expression of DAPG and PLT biosynthetic genes in *P. fluorescens*. Clearly, the plant species has an important role in determining the extent of disease suppression provided by *P. fluorescens* strain CHA0, suggesting that signals present in root exudates may modulate the production of biocontrol factors (Maurhofer et al. 1994, 1995; Schmidli-Sacherer et al. 1997). For instance, *phl* gene expression in strain CHA0 is consistently enhanced in the rhizosphere of cucumber, wheat and maize when *Pythium* infests the roots (Notz et al. 2001). Since neither the fungus itself nor its culture filtrates stimulate DAPG expression in vitro, increased *phl* gene expression is presumably due to alterations in root exudates caused by pathogen attack (Notz et al. 2001; Maurhofer et al. 2002). Furthermore, the *phl* genes are more strongly expressed on wheat and maize roots than on cucumber and bean roots, pointing to differences in root exudate composition between these monocotyledonous and dicotyledonous plants (Notz et al. 2001). However, the signals involved are largely unknown. Possible candidates are phenolic metabolites that are released by the plants as root exudates or as cell wall degradation products. Phenolic compounds have a number of ecological functions, e.g. as growth inhibitors of other plant species, as antimicrobial agents, or as specific signals guiding interactions between plants and rhizobia or *Agrobacterium* spp. (Siqueira et al. 1991; Phillips and Kapulnik 1995; Hirsch et al. 2004; Bais et al. 2006). Among several compounds of plant origin, indole-3-acetic acid (IAA) stimulates *phl* gene expression, resorcinol represses *plt* gene expression and salicylate downregulates the expression of both *phl* and *plt* genes in strain CHA0 (Schnider-Keel et al. 2000; Baehler et al. 2005; de Werra et al. 2006).

Phenazines and AHLs as signals

Phenazine compounds are coloured heterocycles derived from chorismate. They are synthesized by Gram-negative and -positive bacteria. Phenazine-1-carboxylic acid (PCA) is the primary product of a set of enzymes encoded by the *phzABCDEFG* operon in *P. fluorescens* 2-79, *P. chlororaphis* PCL1391, *P. aeruginosa* PAO and *P. aureofaciens* 30-84 (although in the last organism a different nomenclature is used for the *phz* operon) (Pierson et al. 1995; Mavrodi et al. 1998, 2006; McDonald et al. 2001).

Modification reactions lead from PCA to phenazine-1-carboxamide (with *phzH*), 2-hydroxyphenazine-1-carboxylic acid (with *phzO*), 1-hydroxyphenazine (with *phzS*) or pyocyanine (=1-hydroxy-5-methyl-phenazinium betaine, with *phzM* and *phzS*) (Chin-A-Woeng et al. 2001a; Delaney et al. 2001; Mavrodi et al. 2001; Parsons et al. 2007). The first demonstration of PCA being a biocontrol determinant came from a study on the suppression of take-all by *P. fluorescens* 2-79 (Thomashow and Weller 1988); take-all is an important disease of wheat and barley caused by *G. graminis* var. *tritici* (Ggt). Both phenazine-1-carboxamide and 2-hydroxyphenazine-1-carboxylic acid, which have stronger antifungal activity against *Fusarium* spp. and Ggt than has PCA, are particularly important for biocontrol activity of *P. chlororaphis* PCL1391 and *P. aureofaciens* 30-84, respectively (Chin-A-Woeng et al. 2001a; Delaney et al. 2001). A combination of PCA and DAPG produced by a genetically engineered derivative of *P. fluorescens* Q8r1-96 has proved effective against Rhizoctonia root rot of wheat (Huang et al. 2003). We note in passing that certain strains of *P. aeruginosa*, while opportunistic animal and human pathogens, are well adapted to the rhizosphere (Berg et al. 2005) where they can display biocontrol properties (Troxler et al. 1997; Ge et al. 2004) and stimulate induced systemic resistance (Audenaert et al. 2002). In the root-colonizing biocontrol strain *P. aeruginosa* 7NSK2, a combination of pyocyanine and the iron chelator pyochelin induces systemic resistance against the leaf pathogens *Botrytis cinerea* on tomato and *Magnaporte grisea* on rice (Audenaert et al. 2002; de Vleesschauwer et al. 2006).

At the transcriptional level, phenazine biosynthesis is controlled by the PhzR-PhzI quorum sensing system. PhzR is a transcriptional activator of the *phz* operons in *P. fluorescens* 2-79, *P. chlororaphis* PCL1391 and *P. aureofaciens* 30-84 and PhzI is an autoinducer synthase producing mostly *N*-(3-hydroxy-hexanoyl)-homoserine lactone (in strain 2-79) and *N*-hexanoyl-homoserine lactone (in the other two strains). These AHLs activate PhzR (Wood and Pierson 1996; Chancey et al. 1999; Chin-A-Woeng et al. 2001b; Khan et al. 2005). Additional regulators of phenazine production have been identified, e.g. RpeA, a repressor of phenazine production in *P. aureofaciens* 30-84 (Whistler and Pierson 2003), and Pip, which activates phenazine-1-carboxamide production in *P. chlororaphis* PCL1391, together with the stress and stationary phase sigma factor RpoS (Girard et al. 2006). In *P. aeruginosa*, QS regulation of secondary metabolism including phenazine biosynthesis is highly complex, involving a hierarchically organized system with two autoinducers, i.e. *N*-(3-oxododecanoyl)-homoserine lactone (the reaction product of the LasI enzyme and principal activator of the LasR transcription factor) and *N*-butanoyl-homoserine lactone (the reaction product of the RhlI enzyme and principal activator of the RhlR transcription factor), and the *Pseudomonas* quinolone signal (PQS) (Lazdunski et al. 2004; Juhas et al. 2005; Price-Whelan et al. 2006). Despite differences in regulatory elements, the pseudomonads described here all share the principle of QS-dependent control of phenazine biosynthesis (Mavrodi et al. 2006). Phenazines themselves may act as late QS signals in the sense that they regulate several dozens of genes that are not directly related to phenazine biosynthesis, during late growth phases (Dietrich et al. 2006).

AHLs regulate phenazine gene expression on roots (Wood et al. 1997) and can serve as interpopulation signals in the wheat rhizosphere (Pierson et al. 1998). AHL ‘mimics’ extracted from *Medicago truncatula* positively influence the expression of QS reporter constructs (Gao et al. 2003), but it is not known whether these compounds have an impact on biocontrol strains in the rhizosphere. By contrast, interference with AHL signalling has been observed in biocontrol bacteria: the phenazine-producer *P. chlororaphis* PCL1391 lost its ability to protect tomato against Fusarium wilt in the presence of AHL-degrading rhizobacteria (Molina et al. 2003) and unidentified signal molecules from a subpopulation of wheat rhizosphere-associated bacteria were found to affect phenazine gene expression negatively in *P. aureofaciens* 30-84 (Morello et al. 2004). Other interference may come from soil fungi which are a potential reservoir of QS ‘quenchers’. For instance, penicillic acid and patulin are secondary metabolites of *Penicillium* spp. that inhibit the expression of QS reporter constructs (Rasmussen et al. 2005). Fusaric acid, a fungal metabolite, represses phenazine-1-carboxamide biosynthesis and *N*-hexanoyl-homoserine lactone production in *P. chlororaphis* PCL1391 (van Rij et al. 2005). Interestingly, salicylate, a plant defence signal, down-regulates pyocyanine formation by *P. aeruginosa* on plant roots (Prithiviraj et al. 2005).

Signals affecting biocontrol factor expression at a post-transcriptional level

Small RNAs serve multiple regulatory purposes in bacteria, mostly at the level of mRNA stability and translation initiation. In many saprophytic bacteria, QS-dependent gene regulation can involve several small RNAs at critical checkpoints. *Vibrio cholerae* provides an excellent example: in this human pathogen the expression of the central virulence regulator HapR is determined by three converging QS pathways. All together, seven small RNAs are involved in these signal transduction pathways (Lenz et al. 2005). In one branch, the VarS/VarA two-component system positively controls the expression of three small RNAs termed CsrB, CsrC and CsrD. VarS is a sensor protein located in the cytoplasmic membrane and VarA is the cognate response regulator. Interaction of VarS with an external signal is assumed to result in autophosphorylation of VarS and in subsequent phosphotransfer to VarA. The three VarA-dependent small RNAs have a strong affinity for the small dimeric RNA-binding protein CsrA (Lenz et al. 2005), a translational regulator whose structure and function is conserved in a wide range of Gram-negative bacteria (Babitzke and Romeo 2007). Thus, at high cell densities, the small RNAs CsrB-CsrC-CsrD sequester CsrA and this ultimately favours the expression of HapR, resulting in repression of virulence and biofilm genes (Bejerano-Sagie and Xavier 2007).

In fluorescent pseudomonads, as in other Gram-negative bacteria, the same (VarS/VarA-like) QS pathway is conserved although the output can vary. In *P. fluorescens* CHA0, this QS pathway has a major role in positively regulating biocontrol factor expression in various host-pathogen systems (Laville et al. 1992; Siddiqui et al. 2005). It is initiated by the GacS/GacA two-component system, which is homologous to VarS/VarA, and it involves three functional homologs of CsrB-CsrC-CsrD, termed RsmX, RsmY and RsmZ, and two homologs of CsrA, termed RsmA and RsmE (Laville et al. 1992; Zuber et al. 2003; Reimmann et al. 2005; Kay et al. 2005) (Fig. 2a). The importance of the GacS/GacA pathway is supported by studies on *gacS* and *gacA* mutants of *P. fluorescens* BL915 and *Pseudomonas* sp. PCL1171 which have lost biocontrol properties (Ligon et al. 1999; van den Broek et al. 2003).

In strains of *P. aeruginosa*, the conserved GacS/GacA system directs the synthesis of two small RNAs, designated RsmY and RsmZ, which antagonize one CsrA homolog, named RsmA (Reimmann et al. 1997; Pessi et al. 2001; Heurlier et al. 2004; Kay et al. 2006) (Fig. 2b). In both *P. aeruginosa* PAO and *P. fluorescens* CHA0, the sequestration of the RNA-binding proteins (RsmA, and RsmE in the latter bacterium) by the small RNAs leads to translation of target mRNAs. Many target mRNAs specify enzymes for secondary metabolism or lytic exoenzymes. Typical target genes of *P. fluorescens* CHA0 are *hcnA* (for an HCN synthase subunit), *phlA* (for a subunit of the DAPG biosynthetic enzyme complex) and *aprA* (for exoprotease AprA) (Blumer et al. 1999; Zuber et al. 2003; Kay et al. 2005) (Fig. 2a). In *P. aeruginosa*, target genes include *hcnA* and the *phz* genes (Reimmann et al. 1997; Pessi et al. 2001) (Fig. 2b). Thus, there is in fact considerable overlap between virulence factors of *P. aeruginosa* and biocontrol traits of *P. fluorescens*. For instance, HCN, phenazines and alkaline protease qualify for both attributes (Haas et al. 2004).

In *P. aeruginosa*, there are two sensors, RetS and LadS, in addition to GacS, all of which appear to determine the activity of GacA. RetS is an antagonist of GacA and might function as a phosphatase removing the phosphate group from phosphorylated GacA (Goodman et al. 2004; Laskowski and Kazmierczak 2006; Yahr and Wolfgang 2006). LadS, like GacS, appears to activate GacA (Ventre et al. 2006). Stimulation of *rsmY* and *rsmZ* transcription by phosphorylated GacA requires RsmA and probably further unknown factors (Fig. 2b). According to genomic sequence data, RetS and LadS also occur in *P. fluorescens* and *P. putida* strains, but the roles of these sensors remain to be determined. In strain CHA0, both RsmA and RsmE are involved in the expression of RsmX, RsmY and RsmZ (Fig. 2a).

In *P. aeruginosa*, the GacS/GacA system positively regulates the expression of the *lasR*, *rhlR* and *rhlI* genes and thereby stimulates the synthesis of *N*-butanoyl-homoserine lactone (Reimmann et al. 1997; Kay et al. 2006) (Fig. 2b). A similar positive effect of the GacS/GacA system on AHL synthesis is also found in biocontrol strains of *P. chlororaphis*, *P. aureofaciens* and *P. putida* (Chancey et al. 1999; Chin-A-Woeng et al. 2001b; Bertani and Venturi 2004; Han et al. 2006). In all fluorescent pseudomonads examined,

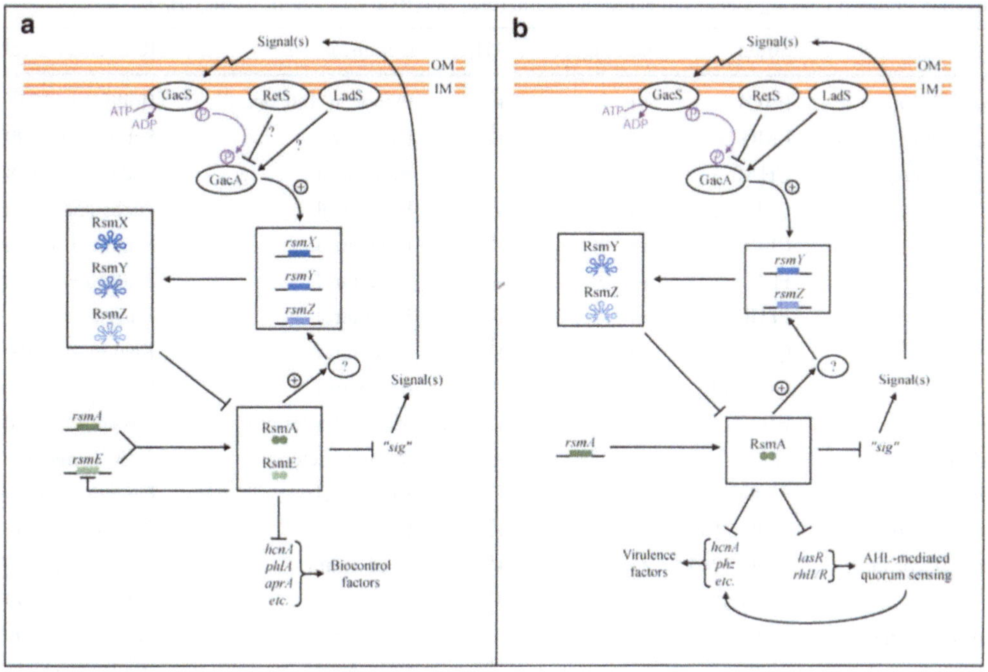

Fig. 2 Post-transcriptional signal transduction pathway downstream of the GacA response regulator in *P. fluorescens* CHA0 (**a**) and *P. aeruginosa* PAO1 (**b**). For a description of the mechanisms, see text. OM, outer membrane; IM, inner membrane; *sig*, hypothetical signal biosynthetic genes. Violet arrows indicate phosphotransfer

GacA function appears to favour the biofilm mode of growth, as opposed to planktonic life of single cells (Heeb and Haas 2001; Goodman et al. 2004; Ventre et al. 2006).

The signal molecules that activate the GacS/GacA cascade in bacteria have not been identified chemically nor are the signal biosynthetic genes known. The signals are unrelated to AHLs and some other well-known QS signals such as PQS, autoinducer 2 (AI-2; a furanosyl diester), or 3-hydroxy-palmitate methyl ester of *Ralstonia* spp. (Dubuis et al. 2006). In both *P. fluorescens* CHA0 and *P. aeruginosa* PAO, the signals activating the GacS/GacA cascade are produced at high cell population densities and under positive GacA control. Mutants that are defective for GacA or the small RNAs (RsmX-RsmY-RsmZ and RsmY-RsmZ, respectively) synthesize $\leq$5% of signal, by comparison with the wild-type (Kay et al. 2005, 2006). Thus, the Gac signals have the characteristics of autoinducers (Fig. 2). In strain CHA0, thiamine is necessary for signal activity, biocontrol factor expression and suppression of *P. ultimum* on cress (Dubuis et al. 2006). Two tentative explanations can be offered for this effect at present. Either signal biosynthesis requires relatively elevated levels of thiamine pyrophosphate as a cofactor such that a limiting thiamine supply will support central metabolism and growth, but not signal production, or thiamine could be a component of the Gac/Rsm signal transduction pathway. There is the intriguing possibility that thiamine triphosphate, a derivative of thiamine pyrophosphate, could be an intracellular signal. In *Escherichia coli*, amino acid starvation leads to transient accumulation of thiamine triphosphate (Lakaye et al. 2004).

AHLs are not the only QS signals that different species or genera of microorganisms can share to establish cross-talk (Riedel et al. 2001). The same has also been observed for AI-2 and for the Gac signals (Bassler et al. 1997; Dubuis and Haas 2007). A survey of beneficial and pathogenic plant-associated pseudomonads reveals that many species produce and release signal molecules activating the Gac/Rsm cascade in *P. fluorescens* and that this signal activity does not correlate with AHL production (Table 1). The bioassay used to detect Gac signals is illustrated in Fig. 3: The expression of the small RNA gene *rsmZ* and that of the biocontrol genes *hcnA* and *phlA*

Table 1 Bacterial strains producing signal molecules activating the Gac/Rsm cascade in strain CHA0

Species	Strain	Gac signal[a]	AHLs	Comments	Reference or source
Pseudomonas aeruginosa	PAO1	++	+[b,c]	Pathogen, biocontrol[e,f]	Holloway (1955)
Pseudomonas corrugata	LMG 2172	++	+[b,c]	Pathogen[f]	Sutra et al. (1997)
Pseudomonas fluorescens	CHA0	++	–	Biocontrol	Stutz et al. (1986)
	CM1′A2	++	–	Biocontrol	Fuchs and Défago (1991)
	F113	++	+[d]	Biocontrol	Fenton et al. (1992)
	Pf-5	++	–	Biocontrol	Paulsen et al. (2005)
	PITR2	++	–	Biocontrol	Harrison et al. (1993)
	Q37-87	++	+[b]	Biocontrol	Keel et al. (1996)
	Q65c-80	++	–	Biocontrol	Harrison et al. (1993)
Vibrio harveyi	BB120	++	+[d]	Marine bacterium	Bassler et al. (1997)
Vibrio natriegens		++	–	Marine bacterium	A. Kukangara
Pseudomonas aureofaciens	30-84	+	+[b,c]	Biocontrol	Pierson et al. (1995)
Pseudomonas caricapapayae	LMG 2152	+	+[b,c]	Pathogen[f]	Sutra et al. (1997)
Pseudomonas chlororaphis	LMG 1245	+	+[b,c]	Type strain	Sutra et al. (1997)
	LMG 5004	+	+[b,c]	Type strain	Sutra et al. (1997)
Pseudomonas fluorescens	LMG 1794	+	–	Type strain	Sutra et al. (1997)
	2-79	+	+[b,c]	Biocontrol	Weller (1983)
	M114	+	–	Biocontrol	Fenton et al. (1992)
	P3	+	+[b]	Soil bacterium	Sharifi-Tehrani et al. (1998)
	P12	+	+[b,c]	Biocontrol	Keel et al. (1996)
	Pf0-1	+	–	Soil bacterium	Compeau et al. (1988)
	PILH1	+	–	Biocontrol	Keel et al. (1996)
	Q69c-80	+	–	Biocontrol	Pierson and Weller (1994)
	SBW25	+	–	Biocontrol	Rainey and Bailey (1996)
Pseudomonas putida	LMG 2257	+	–	Type strain	Sutra et al. (1997)
	KD	+	–	Biocontrol	Sharifi-Tehrani et al. (1998)
	KT2440	+	–	Soil bacterium	Franklin et al. (1981)
Xanthomonas campestris	8004	+	–	Pathogen[f]	Turner et al. (1984)
Escherichia coli	DH5α	–	–	Laboratory strain	Sambrook and Russell (2001)
Pseudomonas alcaligenes	Ps93	–	–	Lipase producing strain	Gerritse et al. (1998)

[a] ++, high signal activity, similar to that of *P. fluorescens* CHA0; +, intermediate signal activity, below that of *P. fluorescens* CHA0; –, no or insignificant signal activity detectable. Extracts containing the signals were obtained from 50-ml cultures as described before (Dubuis et al. 2006) and tested with *hcnA′-′lacZ*, *rsmZ-lacZ* and *phlA′-′lacZ* reporter constructs as shown in Fig. 4. Adapted from Dubuis (2005)

[b] +, Long chain AHLs detected by the reporter strain *Agrobacterium tumefaciens* NTL4/pZLR4 (Shaw et al. 1997)

[c] +, Short chain AHLs detected by the reporter strain *Chromobacterium violaceum* CV026 (McClean et al. 1997)

[d] AHLs produced by *P. fluorescens* F113 (*N*-(3-hydroxy-7-*cis*-tetradecenoyl)-homoserine lactone; Laue et al. 2000 and *V. harveyi* *N*-3-hydroxybutanoyl-homoserine lactone; Bassler et al. 1997) were not detected by the reporter strains *A. tumefaciens* NTL4/pZLR4 and *C. violaceum* CV026

[e] Pathogens for vertebrates

[f] Pathogens for plants

is stimulated about three-fold in *P. fluorescens* CHA0 by saturating amounts of the homologous signals. No stimulation is observed in a *gacA* mutant background (Dubuis and Haas 2007). Signals extracted from culture supernatants of several biocontrol pseudomonads or from the pathogens *P. aeruginosa*, *P. corrug-*

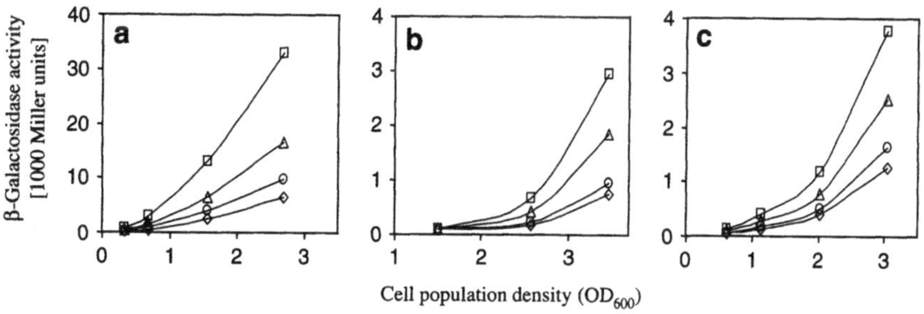

Fig. 3 Bioassay of signal activity. Dichloromethane-extracted supernatants from bacterial strains were tested with *hcnA* (**a**), *rsmZ* (**b**) and *phlA* (**c**) reporter strains as described elsewhere (Dubuis et al. 2006). ß-Galactosidase measurements were carried out in triplicate using cells from the reporter strains *P. fluorescens* CHA0/pME6530 (*hcnA'-'lacZ*; a), CHA0/pME6091 (*rsmZ-lacZ*; b) and CHA0/pME6702 (*phlA'-'lacZ*; c). ◇, control without added extract; ○, plus extract from strain having weak or no signal activity; △, plus extract from strain having intermediate signal activity; □, plus extract from strain having strong, CHA0-like activity; OD_{600}, optical density at 600 nm. Adapted from Dubuis (2005)

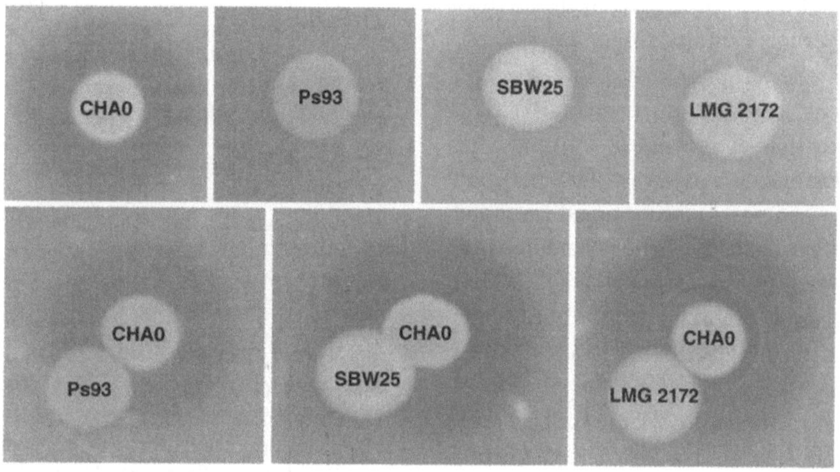

Fig. 4 Antibiotic production by *P. fluorescens* CHA0 in the presence of *P. fluorescens* SBW25, *P. corrugata* LMG 2172 or *P. alcaligenes* Ps93. Strains SBW25, LMG 2172 and Ps93 were inoculated by placing 5-μl drops of overnight cultures onto nutrient agar. After incubation at 30°C for 24 h, 5-μl drops of a CHA0 culture were added and incubation was continued for 24 h. An overlay with *Bacillus subtilis* revealed antibiotic production by growth inhibition zones

ata, *P. caricapapayae* and *Xanthomonas campestris* have similarly high or somewhat lower inducing activities (Table 1; Fig. 3). In this test, a *rpfF* mutant of *X. campestris*, which does not produce the diffusible signal factor (DSF; *cis*-11-methyl-dodecenoate), is still active. Interestingly, two *Vibrio* spp. have been found to have good signal activity, whereas *P. alcaligenes* and *E. coli* DH5α are devoid of it (Dubuis and Haas 2007) (Table 1). Signal activity can be visualized as stimulation of antibiotic production in strain CHA0. When colonies of *P. fluorescens* SBW25 or *P. corrugata* LMG 2172 (both of which do not produce detectable antibiotic compounds) grow next to a colony of strain CHA0, greatly enhanced halos of growth inhibition of a sensitive indicator, *Bacillus subtilis*, can be observed. By contrast, the signal-negative strain Ps93 of *P. alcaligenes* has no effect on antibiotic production (Fig. 4). These experiments demonstrate the ability of various plant-associated bacteria to stimulate post-transcriptionally the expression of biocontrol factors (here: essentially DAPG) in *P. fluorescens* CHA0. Although inhibitory compounds would also be detected by the bioassay shown in Fig. 3, none of the microorganisms tested in Table 1 appears to have such activity.

Conclusions

Biocontrol of root diseases by fluorescent pseudomonads is multifactorial; important mechanisms include antagonism between the biocontrol agents and the pathogens, degradation of virulence factors and induction of systemic resistance in the plant. Not surprisingly, both the biocontrol agents and the pathogens have developed defence mechanisms to minimize the impact of antagonism and some of these strategies have been shown to be active in the rhizosphere. Unlike predation, the biocontrol strategies that we have reviewed here work at a distance (of several micrometers) and rely on the emission or destruction of chemical signals. Root exudates critically influence the outcome, via signals that are mostly unknown and merit to be investigated. However, not all biocontrol interactions are of the antagonistic kind. Some previously unsuspected, positive interactions between rhizosphere bacteria stimulate the expression of biocontrol traits via the Gac/Rsm signal transduction pathway. This pathway operates essentially at a post-transcriptional level and in this respect differs from the better-characterized, AHL-dependent signal transduction pathways, whose outputs manifest themselves at the level of transcription. When activated, the Gac/Rsm cascade favours the biofilm mode of growth and the expression of multiple biocontrol factors in root-colonizing bacteria. Available reporter constructs facilitate the detection of activating signal molecules. In future research, it will be interesting to investigate which factors prevailing in the rhizosphere influence the activity of this signal transduction pathway as well as that of other key regulatory elements in plant-beneficial rhizobacteria.

Acknowledgements We gratefully acknowledge financial support from the Swiss National Science Foundation (projects 3100A0-100180 and 3100A0-105881) and from the State Secretariat for Education and Research (COST project C04.0201).

References

Abbas, A., McGuire, J. E., Crowley, D., Baysse, C., Dow, M., & O'Gara, F. (2004). The putative permease PhlE of *Pseudomonas fluorescens* F113 has a role in 2,4-diacetylphloroglucinol resistance and in general stress tolerance. *Microbiology, 150*, 2443–2450.

Achkar, J., Xian, M., Zhao, H., & Frost, J. W. (2005). Biosynthesis of phloroglucinol. *Journal of the American Chemical Society, 127*, 5332–5333.

Andersen, J. B., Koch, B., Nielsen, T. H., Sørensen, D., Hansen, M., Nybroe, O., Christophersen, C., Sørensen, J., Molin, S., & Givskov, M. (2003). Surface motility in *Pseudomonas* sp. DSS73 is required for efficient biological containment of the root-pathogenic microfungi *Rhizoctonia solani* and *Pythium ultimum. Microbiology, 149*, 1147–1156.

Audenaert, K., Pattery, T., Cornelis, P., & Höfte, M. (2002). Induction of systemic resistance to *Botrytis cinerea* in tomato by *Pseudomonas aeruginosa* 7NSK2: Role of salicylic acid, pyochelin, and pyocyanin. *Molecular Plant-Microbe Interactions, 15*, 1147–1156.

Babitzke, P., & Romeo, T. (2007). CsrB sRNA family: Sequestration of RNA-binding regulatory proteins. *Current Opinion in Microbiology, 10*, 156–163.

Baehler, E., Bottiglieri, M., Péchy-Tarr, M., Maurhofer, M., & Keel, C. (2005). Use of green fluorescent protein-based reporters to monitor balanced production of antifungal compounds in the biocontrol agent *Pseudomonas fluorescens* CHA0. *Journal of Applied Microbiology, 99*, 24–38.

Baehler, E., de Werra, P., Wick, L. Y., Péchy-Tarr, M., Mathys, S., Maurhofer, M., & Keel, C. (2006). Two novel MvaT-like global regulators control exoproduct formation and biocontrol activity in root-associated *Pseudomonas fluorescens* CHA0. *Molecular Plant-Microbe Interactions, 19*, 313–329.

Bais, H. P., Weir, T. L., Perry, L. G., Gilroy, S., & Vivanco, J. M. (2006). The role of root exudates in rhizosphere interactions with plants and other organisms. *Annual Review of Plant Biology, 57*, 233–266.

Bangera, M. G., & Thomashow, L. S. (1999). Identification and characterization of a gene cluster for synthesis of the polyketide antibiotic 2,4-diacetylphloroglucinol from *Pseudomonas fluorescens* Q2-87. *Journal of Bacteriology, 181*, 3155–3163.

Bassler, B. L., Greenberg, E. P., & Stevens, A. M. (1997). Cross-species induction of luminescence in the quorum-sensing bacterium *Vibrio harveyi. Journal of Bacteriology, 179*, 4043–4045.

Bejerano-Sagie, M., & Xavier, K. B. (2007). The role of small RNAs in quorum sensing. *Current Opinion in Microbiology, 10*, 189–198.

Berg, G., Eberl, L., & Hartmann, A. (2005). The rhizosphere as a reservoir for opportunistic human pathogenic bacteria. *Environmental Microbiology, 7*, 1673–1685.

Bertani, I., & Venturi, V. (2004). Regulation of the *N*-acyl homoserine lactone-dependent quorum-sensing system in rhizosphere *Pseudomonas putida* WCS358 and cross-talk with the stationary-phase RpoS sigma factor and the global regulator GacA. *Applied and Environmental Microbiology, 70*, 5493–5502.

Blumer, C., & Haas, D. (2000). Mechanism, regulation, and ecological role of bacterial cyanide biosynthesis. *Archives of Microbiology, 173*, 170–177.

Blumer, C., Heeb, S., Pessi, G., & Haas, D. (1999). Global GacA-steered control of cyanide and exoprotease production in Pseudomonas fluorescens involves specific

ribosome binding sites. *Proceedings of the National Academy of Sciences USA, 96*, 14073–14078.
Bottiglieri, M., & Keel, C. (2006). Characterization of PhlG, a hydrolase that specifically degrades the antifungal compound 2,4-diacetylphloroglucinol in the biocontrol agent *Pseudomonas fluorescens* CHA0. *Applied and Environmental Microbiology, 72*, 418–427.
Brodhagen, M., Henkels, M. D., & Loper, J. E. (2004). Positive autoregulation and signaling properties of pyoluteorin, an antibiotic produced by the biological control organism *Pseudomonas fluorescens* Pf-5. *Applied and Environmental Microbiology, 70*, 1758–1766.
Brodhagen, M., Paulsen, I., & Loper, J. E. (2005). Reciprocal regulation of pyoluteorin production with membrane transporter gene expression in *Pseudomonas fluorescens* Pf-5. *Applied and Environmental Microbiology, 71*, 6900–6909.
Chancey, S. T., Wood, D. W., & Pierson, L. S. 3rd. (1999). Two-component transcriptional regulation of *N*-acyl-homoserine lactone production in *Pseudomonas aureofaciens*. *Applied and Environmental Microbiology, 65*, 2294–2299.
Chin-A-Woeng, T. F. C., Bloemberg, G. V., & Lugtenberg, B. J. J. (2003). Phenazines and their role in biocontrol by *Pseudomonas* bacteria. *New Phytologist, 157*, 503–523.
Chin-A-Woeng, T. F. C., Thomas-Oates, J. E., Lugtenberg, B. J. J., & Bloemberg, G. V. (2001a). Introduction of the *phzH* gene of *Pseudomonas chlororaphis* PCL1391 extends the range of biocontrol ability of phenazine-1-carboxylic acid-producing *Pseudomonas* spp. strains. *Molecular Plant-Microbe Interactions, 14*, 1006–1015.
Chin-A-Woeng, T. F. C., van den Broek, D., de Voer, G., van der Drift, K., Tuinman, S., Thomas-Oates, J. E., Lugtenberg, B. J. J., & Bloemberg, G. V. (2001b). Phenazine-1-carboxamide production in the biocontrol strain *Pseudomonas chlororaphis* PCL1391 is regulated by multiple factors secreted into the growth medium. *Molecular Plant-Microbe Interactions, 14*, 969–979.
Compant, S., Duffy, B., Nowak, J., Clement, C., & Barka, E. A. (2005). Use of plant growth-promoting bacteria for biocontrol of plant diseases: Principles, mechanisms of action, and future prospects. *Applied and Environmental Microbiology, 71*, 4951–4959.
Compeau, G., Al-Achi, B. J., Platsouka, E., & Levy, S. B. (1988). Survival of rifampin-resistant mutants of *Pseudomonas fluorescens* and *Pseudomonas putida* in soil systems. *Applied and Environmental Microbiology, 54*, 2432–2438.
de Souza, J. T., Arnould, C., Deulvot, C., Lemanceau, P., Gianinazzi-Pearson, V., & Raaijmakers, J. M. (2003b). Effect of 2,4-diacetylphloroglucinol on *Pythium*: Cellular responses and variation in sensitivity among propagules and species. *Phytopathology, 93*, 966–975.
de Souza, J. T., Weller, D. M., & Raaijmakers, J. M. (2003a). Frequency, diversity, and activity of 2,4-diacetylphloroglucinol-producing fluorescent *Pseudomonas* spp. in Dutch take-all decline soils. *Phytopathology, 93*, 54–63.
de Vleesschauwer, D., Cornelis, P., & Höfte, M. (2006). Redox-active pyocyanin secreted by *Pseudomonas aeruginosa* 7HSK2 triggers systemic resistance to *Magnaporthe grisea* but enhances *Rhizoctonia solani* susceptibility in rice. *Molecular Plant-Microbe Interactions, 19*, 1406–1419.
de Werra, P., Huser, A., Baehler, E., Keel, C., & Maurhofer, M. (2006). Using flow cytometry for *in situ* monitoring of antimicrobial compound production in the biocontrol bacterium *Pseudomonas fluorescens* CHA0. *IOBC/WPRS Bulletin, 29*, 117–121.
Delaney, S. M., Mavrodi, D. V., Bonsall, R. F., & Thomashow, L. S. (2001). *phzO*, a gene for biosynthesis of 2-hydroxylated phenazine compounds in *Pseudomonas aureofaciens* 30-84. *Journal of Bacteriology, 183*, 318–327.
Delany, I., Sheehan, M. M., Fenton, A., Bardin, S., Aarons, S., & O'Gara, F. (2000). Regulation of production of the antifungal metabolite 2,4-diacetylphloroglucinol in *Pseudomonas fluorescens* F113: Genetic analysis of *phlF* as a transcriptional repressor. *Microbiology, 146*, 537–543.
Dietrich, L. E. P., Price-Whelan, A., Petersen, A., Whiteley, M., & Newman, D. K. (2006). The phenazine pyocyanin is a terminal signalling factor in the quorum sensing network of *Pseudomonas aeruginosa*. *Molecular Microbiology, 61*, 1308–1321.
Dubuis, C. (2005). Cell-cell communication in the biocontrol strain *Pseudomonas fluorescens* CHA0. PhD thesis, Switzerland, Dépatement de microbiologie fondamentale, University of Lausanne.
Dubuis, C., & Haas, D. (2007). Cross-species GacA-controlled induction of antibiosis in pseudomonads. *Applied and Environmental Microbiology, 73*, 650–654.
Dubuis, C., Rolli, J., Lutz, M., Défago, G., & Haas, D. (2006). Thiamine-auxotrophic mutants of *Pseudomonas fluorescens* CHA0 are defective in cell-cell signaling and biocontrol factor expression. *Applied and Environmental Microbiology, 72*, 2606–2613.
Duffy, B., & Défago, G. (1997). Zinc improves biocontrol of fusarium crown and root rot of tomato by *Pseudomonas fluorescens* and represses the production of pathogen metabolites inhibitory to bacterial antibiotic biosynthesis. *Phytopathology, 87*, 1250–1257.
Duffy, B., Keel, C., & Défago, G. (2004). Potential role of pathogen signaling in multitrophic plant-microbe interactions involved in disease protection. *Applied and Environmental Microbiology, 70*, 1836–1842.
Duffy, B., Schouten, A., & Raaijmakers, J. M. (2003). Pathogen self-defense: Mechanisms to counteract microbial antagonism. *Annual Review of Phytopathology, 41*, 501–538.
Fenton, A. M., Stephens, P. M., Crowley, J., O'Callaghan, M., & O'Gara, F. (1992). Exploitation of gene(s) involved in 2,4-diacetylphloroglucinol biosynthesis to confer a new biocontrol capability to a *Pseudomonas* strain. *Applied and Environmental Microbiology, 58*, 3873–3878.
Franklin, F. C., Bagdasarian, M., Bagdasarian, M. M., & Timmis, K. N. (1981). Molecular and functional analysis of the TOL plasmid pWWO from *Pseudomonas putida* and cloning of genes for the entire regulated aromatic ring meta cleavage pathway. *Proceedings of the National Academy of Sciences USA, 78*, 7458–7462.
Fuchs, J., & Défago, G. (1991). Protection of cucumber plants against black root rot caused by *Phomopsis sclerotioides* with rhizobacteria. In C. Keel, B. Koller, & G. Défago (Eds.), *Plant growth-promoting rhizobacteria—progress*

and prospects (pp. 51–56). Interlaken, Switzerland: IOBC/WPRS.

Fuqua, W. C., Winans, S. C., & Greenberg, E. P. (1994). Quorum sensing in bacteria: The LuxR-LuxI family of cell density-responsive transcriptional regulators. *Journal of Bacteriology, 176*, 269–275.

Gantner, S., Schmid, M., Dürr, C., Schuhegger, R., Steidle, A., Hutzler, P., Langebartels, C., Eberl, L., Hartmann, A., & Dazzo, F. B. (2006). *In situ* quantitation of the spatial scale of calling distances and population density-independent *N*-acylhomoserine lactone-mediated communication by rhizobacteria colonized on plant roots. *FEMS Microbiology Ecology, 56*, 188–194.

Gao, M., Teplitski, M., Robinson, J. B., & Bauer, W. D. (2003). Production of substances by *Medicago trucatula* that affect bacterial quorum sensing. *Molecular Plant-Microbe Interactions, 16*, 827–834.

Ge, Y., Huang, X., Wang, S., Zhang, X., & Xu, Y. (2004). Phenazine-1-carboxylic acid is negatively regulated and pyoluteorin positively regulated by GacA in *Pseudomonas* sp. M18. *FEMS Microbiology Letters, 237*, 41–47.

Gerritse, G., Hommes, R. W., & Quax, W. J. (1998). Development of a lipase fermentation process that uses a recombinant *Pseudomonas alcaligenes* strain. *Applied and Environmental Microbiology, 64*, 2644–2651.

Girard, G., Barends, S., Rigali, S., van Rij, E. T., Lugtenberg, B. J. J., & Bloemberg, G. V. (2006). Pip, a novel activator of phenazine biosynthesis in *Pseudomonas chlororaphis* PCL1391. *Journal of Bacteriology, 188*, 8283–8293.

González, J. E., & Keshavan, N. D. (2006). Messing with bacterial quorum sensing. *Microbiology and Molecular Biology Reviews, 70*, 859–875.

Goodman, A. L., Kulasekara, B., Rietsch, A., Boyd, D., Smith, R. S., & Lory, S. (2004). A signaling network reciprocally regulates genes associated with acute infection and chronic persistence in *Pseudomonas aeruginosa*. *Developmental Cell, 7*, 745–754.

Haas, D., Blumer, C., & Keel, C. (2000). Biocontrol ability of fluorescent pseudomonads genetically dissected: Importance of positive feedback regulation. *Current Opinion in Biotechnology, 11*, 290–297.

Haas, D., & Défago, G. (2005). Biological control of soil-borne pathogens by fluorescent pseudomonads. *Nature Reviews Microbiology, 3*, 307–319.

Haas, D., & Keel, C. (2003). Regulation of antibiotic production in root-colonizing *Pseudomonas* spp. and relevance for biological control of plant disease. *Annual Review of Phytopathology, 41*, 117–153.

Haas D., Reimmann C., & Valverde, C. (2004). Common mechanisms in beneficial, deleterious host-microbe interactions. In L. Tikhonovich, B. Lugtenberg, & N. Provorov (Eds.), *Biology of plant-microbe interactions* (Vol. 4, pp. 537–541). St. Paul, Minnesota, USA: International Society for Molecular Plant-Microbe Interactions.

Han, S. H., Lee, S. J., Moon, J. H., Park, K. H., Yang, K. Y., Cho, B. H., Kim, K. Y., Kim, Y. W., Lee, M. C., Anderson, A. J., & Kim, Y. C. (2006). GacS-dependent production of 2R, 3R-butanediol by *Pseudomonas chlororaphis* O6 is a major determinant for eliciting systemic resistance against *Erwinia carotovora* but not against *Pseudomonas syringae* pv. *tabaci* in tobacco. *Molecular Plant-Microbe Interactions, 19*, 924–930.

Harrison, L. A., Letendre, L., Kovacevich, P., Pierson, E., & Weller, D. (1993). Purification of an antibiotic effective against *Gaeumannomyces graminis* var. *tritici* produced by a biocontrol agent, *Pseudomonas aureofaciens*. *Soil Biology and Biochemistry, 25*, 215–221.

Heeb, S., & Haas, D. (2001). Regulatory roles of the GacS/GacA two-component system in plant-associated and other Gram-negative bacteria. *Molecular Plant-Microbe Interactions, 14*, 1351–1363.

Hense, B. A., Kuttler, C., Müller, J., Rothballer, M., Hartmann, A., & Kreft, J.-U. (2007). Does efficiency sensing unify diffusion and quorum sensing? *Nature Reviews Microbiology, 5*, 230–239.

Hernandez, M. E., Kappler, A., & Newman, D. K. (2004). Phenazines and other redox-active antibioticx promote microbioal mineral reduction. *Applied and Environmental Microbiology, 70*, 921–928.

Heurlier, K., Williams, F., Heeb, S., Dormond, C., Pessi, G., Singer, D., Camara, M., Williams, P., & Haas, D. (2004). Positive control of swarming, rhamnolipid synthesis, and lipase production by the posttranscriptional RsmA/RsmZ system in *Pseudomonas aeruginosa* PAO1. *Journal of Bacteriology, 186*, 2936–2945.

Hirsch, A. M., Bauer, W. D., Bird, D. M., Cullimore, J., Tyler, A. B., & Yoderf, J. I. (2004). Molecular signals and receptors: Controlling rhizosphere interactions between plants and other organisms. *Ecology, 84*, 858–868.

Holloway, B. W. (1955). Genetic recombination in *Pseudomonas aeruginosa*. *Journal of General Microbiology, 13*, 572–581.

Howell, C. R., & Stipanovic, R. D. (1980). Suppression of *Pythium ultimum*-induced damping-off of cotton seedlings by *Pseudomonas fluorescens* and its antibiotic, pyoluteorin. *Phytopathology, 70*, 712–715.

Huang, Z., Bonsall, R. F., Mavrodi, D. V., Weller, D., & Thomashow, L. S. (2003). Transformation of *Pseudomonas fluorescens* with genes for biosynthesis of phenazine-1-carboxylic acid improves biocontrol of rhizoctonia root rot and in situ antibiotic production. *FEMS Microbiology Ecology, 49*, 243–251.

Huang, J. J., Petersen, A., Whiteley, M., & Leadbetter, J. R. (2006a). Identification of QuiP, the product of gene PA1032, as the second acyl-homoserine lactone acylase of *Pseudomonas aeruginosa* PAO1. *Applied and Environmental Microbiology, 72*, 1190–1197.

Huang, X., Yan, A., Zhang, X., & Xu, Y. (2006b). Identification and characterization of a putative ABC transporter PltHIJKN required for pyoluteorin production in *Pseudomonas* sp. M18. *Gene, 376*, 68–78.

Huang, X., Zhu, D., Ge, Y., Hu, H., Zhang, X., & Xu, Y. (2004). Identification and characterization of *pltZ*, a gene involved in the repression of pyoluteorin biosynthesis in *Pseudomonas* sp. M18. *FEMS Microbiology Letters, 232*, 197–202.

Iavicoli, A., Boutet, E., Buchala, A., & Métraux, J. P. (2003). Induced systemic resistance in *Arabidopsis thaliana* in response to root inoculation with *Pseudomonas fluorescens* CHA0. *Molecular Plant-Microbe Interactions, 16*, 851–858.

Johansen, J. E., Binnerup, S. J., Lejbølle, K. B., Mascher, F., Sørensen, J., & Keel, C. (2002). Impact of biocontrol strain *Pseudomonas fluorescens* CHA0 on rhizosphere bacteria isolated from barley (*Hordeum vulgare* L.) with special reference to Cytophaga-like bacteria. *Journal of Applied Microbiology, 93*, 1065–1074.

Jousset, A., Lara, E., Wall, L. G., & Valverde, C. (2006). Secondary metabolites help biocontrol strain *Pseudomonas fluorescens* CHA0 to escape protozoan grazing. *Applied and Environmental Microbiology, 72*, 7083–7090.

Juhas, M., Eberl, L., & Tümmler, B. (2005). Quorum sensing: The power of cooperation in the world of *Pseudomonas*. *Environmental Microbiology, 7*, 459–471.

Kaur, R., Macleod, J., Foley, W., & Nayudu, M. (2006). Gluconic acid: An antifungal agent produced by *Pseudomonas* species in biological control of take-all. *Phytochemistry, 67*, 595–604.

Kay, E., Dubuis, C., & Haas, D. (2005). Three small RNAs jointly ensure secondary metabolism and biocontrol in *Pseudomonas fluorescens* CHA0. *Proceedings of the National Academy of Sciences USA,* 102, 17136–17141.

Kay, E., Humair, B., Dénervaud, V., Riedel, K., Spahr, S., Eberl, L., Valverde, C., & Haas, D. (2006). Two GacA-dependent small RNAs modulate the quorum sensing response in *Pseudomonas aeruginosa*. *Journal of Bacteriology, 188*, 6026–6033.

Keel, C., Schnider, U., Maurhofer, M., Voisard, C., Laville, J., Burger, U., Wirthner, P., Haas, D., & Défago, G. (1992). Suppression of root diseases by *Pseudomonas fluorescens* CHA0: Importance of the bacterial secondary metabolite 2,4-diacetylphloroglucinol. *Molecular Plant-Microbe Interactions, 5*, 4–13.

Keel, C., Weller, D. M., Natsch, A., Défago, G., Cook, R. J., & Thomashow, L. S. (1996). Conservation of the 2,4-diacetylphloroglucinol biosynthesis locus among fluorescent *Pseudomonas* strains from diverse geographic locations. *Applied and Environmental Microbiology, 62*, 552–563.

Keel, C., Wirthner, P., Oberhänsli, T., Voisard, C., Burger, U., Haas, D., & Défago, G. (1990). Pseudomonads as antagonists of plant pathogens in the rhizosphere: Role of the antibiotic 2,4-diacetylphloroglucinol in the suppression of black root rot of tobacco. *Symbiosis, 9*, 327–341.

Khan, S. R., Mavrodi, D. V., Jog, G. J., Suga, H., Thomashow, L. S., & Farrand, S. K. (2005). Activation of the *phz* operon of *Pseudomonas fluorescens* 2-79 requires the LuxR homolog PhzR, *N*-(3-OH-hexanoyl)-l-homoserine lactone produced by the LuxI homolog PhzI, and a cis-acting *phz* box. *Journal of Bacteriology, 187*, 6517–6527.

Kraus, J., & Loper, J. E. (1995). Characterization of a genomic region required for production of the antibiotic pyoluteorin by the biological control agent *Pseudomonas fluorescens* Pf-5. *Applied and Environmental Microbiology, 61*, 849–854.

Kuiper, I., Lagendijk, E. L., Pickford, R., Derrick, J. P., Lamers, G. E., Thomas-Oates, J. E., Lugtenberg, B. J. J., & Bloemberg, G. V. (2004). Characterization of two *Pseudomonas putida* lipopeptide biosurfactants, putisolvin I and II, which inhibit biofilm formation and break down existing biofilms. *Molecular Microbiology, 51*, 97–113.

Lakaye, B., Wirtzfeld, B., Wins, P., Grisar, T., & Bettendorff, L. (2004). Thiamine triphosphate, a new signal required for optimal growth of *Escherichia coli* during amino acid starvation. *Journal of Biological Chemistry, 279*, 17142–17147.

Laskowski, M. A., & Kazmierczak, B. I. (2006). Mutational analysis of RetS, an unusual sensor kinase-response regulator hybrid required for *Pseudomonas aeruginosa* virulence. *Infection and Immunity, 74*, 4462–4473.

Laue, B. E., Jiang, Y., Chhabra, S. R., Jacob, S., Stewart, G. S., Hardman, A., Downie, J. A., O'Gara, F., & Williams, P. (2000). The biocontrol strain *Pseudomonas fluorescens* F113 produces the Rhizobium small bacteriocin, *N*-(3-hydroxy-7-cis-tetradecenoyl)-homoserine lactone, via HdtS, a putative novel *N*-acylhomoserine lactone synthase. *Microbiology, 146*, 2469–2480.

Laville, J., Voisard, C., Keel, C., Maurhofer, M., Défago, G., & Haas, D. (1992). Global control in *Pseudomonas fluorescens* mediating antibiotic synthesis and suppression of black root rot of tobacco. *Proceedings of the National Academy of Sciences USA, 89*, 1562–1566.

Lazdunski, A. M., Ventre, I., & Sturgis, J. N. (2004). Regulatory circuits and communication in Gram-negative bacteria. *Nature Reviews Microbiology, 2*, 581–592.

Lenz, D. H., Miller, M. B., Zhu, J., Kulkarni, R. V., & Bassler, B. L. (2005). CsrA and three redundant small RNAs regulate quorum sensing in *Vibrio cholerae*. *Molecular Microbiology, 58*, 1186–1202.

Ligon, J. M., Hill, D. S., Hammer, P. E., & Torkewitz, N. R. (1999). Genetic modifications of *Pseudomonas* that enhance biological disease control. *Acta Horticulturae, 504*, 53–60.

Lugtenberg, B. J. J., & Bloemberg, G. V. (2004). Life in the rhizosphere. In J. L. Ramos (Ed.), *Pseudomonas: Genomics, life style and molecular archtecture* (Vol. 1, pp. 403–430). New York, NY, USA: Kluwer Academic/Plenum Publishers.

Lugtenberg, B. J. J., Dekkers, L., & Bloemberg, G. V. (2001). Molecular determinants of rhizosphere colonization by *Pseudomonas*. *Annual Review of Phytopathology, 39*, 461–490.

Lutz, M., Wenger, S., Maurhofer, M., Défago, G., & Duffy, B. (2004). Signaling between bacterial and fungal biocontrol agents in a strain mixture. *FEMS Microbiology Ecology, 48*, 447–455.

Maurhofer, M., Baehler, E., Notz, R., Martinez, V., & Keel, C. (2004). Cross talk between 2,4-diacetylphloroglucinol-producing biocontrol pseudomonads on wheat roots. *Applied and Environmental Microbiology, 70*, 1990–1998.

Maurhofer, M., Dubach, H., Haas, D., & Défago, G. (2002). How soilborne fungal pathogens affect the production of 2,4-diacetylphloroglucinol in biocontrol strain *Pseudomonas fluorescens* CHA0. *IOBC/WPRS Bulletin, 25*, 102–106.

Maurhofer, M., Keel, C., Haas, D., & Défago, G. (1994). Pyoluteorin production by *Pseudomonas fluorescens* strain CHA0 is involved in the suppression of *Pythium* damping-off of cress but not of cucumber. *European Journal of Plant Pathology, 100*, 221–232.

Maurhofer, M., Keel, C., Haas, D., & Défago, G. (1995). Influence of plant species on disease suppression by *Pseudomonas fluorescens* strain CHA0 with enhanced antibiotic production. *Plant Pathology, 44*, 40–50.

Maurhofer, M., Keel, C., Schnider, U., Voisard, C., Haas, D., & Défago, G. (1992). Influence of enhanced antibiotic production in *Pseudomonas fluorescens* strain CHA0 on its disease suppressive capacity. *Phytopathology, 82*, 190–195.

Mavrodi, D. V., Blankenfeldt, W., & Thomashow, L. S. (2006). Phenazine compounds in fluorescent *Pseudomonas* spp. biosynthesis and regulation. *Annual Review of Phytopathology, 44*, 417–445.

Mavrodi, D. V., Bonsall, R. F., Delaney, S. M., Soule, M. J., Phillips, G., & Thomashow, L. S. (2001). Functional analysis of genes for biosynthesis of pyocyanin and phenazine-1-carboxamide from *Pseudomonas aeruginosa* PAO1. *Journal of Bacteriology, 183*, 6454–6565.

Mavrodi, D. V., Ksenzenko, V. N., Bonsall, R. F., Cook, R. J., Boronin, A. M., & Thomashow, L. S. (1998). A seven-gene locus for synthesis is of phenazine-1-carboxylic acid by *Pseudomonas fluorescens* 2-79. *Journal of Bacteriology, 180*, 2541–2548.

Mazzola, M., Fujimoto, D. K., Thomashow, L. S., & Cook, R. J. (1995). Variation in sensitivity of *Gaeumannomyces graminis* to antibiotics produced by fluorescent *Pseudomonas* spp. and effect on biological control of take-all of wheat. *Applied and Environmental Microbiology, 61*, 2554–2559.

McClean, K. H., Winson, M. K., Fish, L., Taylor, A., Chhabra, S. R., Camara, M., Daykin, M., Lamb, J. H., Swift, S., Bycroft, B. W., Stewart, G. S., & Williams, P. (1997). Quorum sensing and *Chromobacterium violaceum*: Exploitation of violacein production and inhibition for the detection of *N*-acylhomoserine lactones. *Microbiology, 143*, 3703–3711.

McDonald, M., Mavrodi, D. V., Thomashow, L. S., & Floss, H. G. (2001). Phenazine biosynthesis in *Pseudomonas fluorescens*: Branchpoint from the primary shikimate biosynthetic pathway and role of phenazine-1,6-dicarboxylic acid. *Journal of the American Chemical Society, 123*, 9459–9460.

Moënne-Loccoz, Y., & Défago, G. (2004). Life as a biocontrol pseudomonad. In J. L. Ramos (Ed.), *Pseudomonas: Genomics, life style and molecular archtecture* (Vol. 1, pp. 457–476). New York, NY, USA: Kluwer Academic/Plenum Publishers.

Molina, L., Constantinescu, F., Michel, L., Reimmann, C., Duffy, B., & Défago, G. (2003). Degradation of pathogen quorum-sensing molecules by soil bacteria: A preventive and curative biological control mechanism. *FEMS Microbiology Ecology, 45*, 71–81.

Morello, J. E., Pierson, E. A., & Pierson, L. S. 3rd. (2004). Negative cross-communication among wheat rhizosphere bacteria: Effect on antibiotic production by the biological control bacterium *Pseudomonas aureofaciens* 30-84. *Applied and Environmental Microbiology, 70*, 3103–3109.

Nealson, K. H. (1977). Autoinduction of bacterial luciferase. Occurrence, mechanism and significance. *Archives of Microbiology, 112*, 73–79.

Notz, R., Maurhofer, M., Dubach, H., Haas, D., & Défago, G. (2002). Fusaric acid-producing strains of *Fusarium oxysporum* alter 2,4-diacetylphloroglucinol biosynthetic gene expression in *Pseudomonas fluorescens* CHA0 in vitro and in the rhizosphere of wheat. *Applied and Environmental Microbiology, 68*, 2229–2235.

Notz, R., Maurhofer, M., Schnider-Keel, U., Duffy, B., Haas, D., & Défago, G. (2001). Biotic factors affecting expression of the 2,4-diacetylphloroglucinol biosynthesis gene *phlA* in *Pseudomonas fluorescens* biocontrol strain CHA0 in the rhizosphere. *Phytopathology, 91*, 873–881.

Nowak-Thompson, B., Chaney, N., Wing, J. S., Gould, S. J., & Loper, J. E. (1999). Characterization of the pyoluteorin biosynthetic gene cluster of *Pseudomonas fluorescens* Pf-5. *Journal of Bacteriology, 181*, 2166–2174.

Parsons, J. F., Greenhagen, B. T., Shi, K., Calabrese, K., Robinson, H., & Ladner, J. E. (2007). Structural and functional analysis of the pyocyanin biosynthetic protein PhzM from *Pseudomonas aeruginosa*. *Biochemistry, 46*, 1821–1828.

Paulsen, I. T., Press, C. M., Ravel, J., Kobayashi, D. Y., Myers, G. S., Mavrodi, D. V., DeBoy, R. T., Seshadri, R., Ren, Q., Madupu, R., Dodson, R. J., Durkin, A. S., Brinkac, L. M., Daugherty, S. C., Sullivan, S. A., Rosovitz, M. J., Gwinn, M. L., Zhou, L., Schneider, D. J., Cartinhour, S. W., Nelson, W. C., Weidman, J., Watkins, K., Tran, K., Khouri, H., Pierson, E. A., Pierson, L. S. 3rd, Thomashow, L. S., & Loper, J. E. (2005). Complete genome sequence of the plant commensal *Pseudomonas fluorescens* Pf-5. *Nature Biotechnology, 23*, 873–878.

Péchy-Tarr, M., Bottiglieri, M., Mathys, S., Bang-Lejbølle, K., Schnider-Keel, U., Maurhofer, M., & Keel, C. (2005). RpoN (sigma^{54}) controls production of antifungal compounds and biocontrol activity in *Pseudomonas fluorescens* CHA0. *Molecular Plant-Microbe Interactions, 18*, 260–272.

Pessi, G., Williams, F., Hindle, Z., Heurlier, K., Holden, M. T., Camara, M., Haas, D., & Williams, P. (2001). The global posttranscriptional regulator RsmA modulates production of virulence determinants and *N*-acylhomoserine lactones in *Pseudomonas aeruginosa*. *Journal of Bacteriology, 183*, 6676–6683.

Phillips, D. A., Fox, T. C., King, M. D., Bhuvaneswari, T. V., & Teuber, L. R. (2004). Microbial products trigger amino acid exudation from plant roots. *Plant Physiology, 136*, 2887–2894.

Phillips, D. A., & Kapulnik, Y. (1995). Plant isoflavonoids, pathogens and symbionts. *Trends in Microbiology, 3*, 58–64.

Pierson, L. S. 3rd, Gaffney, T., Lam, S., & Gong, F. (1995). Molecular analysis of genes encoding phenazine biosynthesis in the biological control bacterium. *Pseudomonas aureofaciens* 30-84. *FEMS Microbiology Letters, 134*, 299–307.

Pierson, E. A., & Weller, D. M. (1994). Use of mixtures of fluorescent pseudomonads to suppress take-all and improve the growth of wheat. *Phytopathology, 84*, 940–947.

Pierson, L. S. 3rd, Wood, D. W., & Pierson, E. A. (1998). Homoserine lactone-mediated gene regulation in plant-associated bacteria. *Annual Review of Phytopathology, 36*, 207–225.

Price-Whelan, A., Dietrich, L. E., & Newman, D. K. (2006). Rethinking 'secondary' metabolism: Physiological roles for phenazine antibiotics. *Nature Chemical Biology, 2*, 71–78.

Prithiviraj, B., Bais, H. P., Weir, T., Suresh, B., Najarro, E. H., Dayakar, B. V., Schweizer, H. P., & Vivanco, J. M. (2005). Down regulation of virulence factors of *Pseudomonas aeruginosa* by salicylic acid attenuates its virulence on *Arabidopsis thaliana* and *Caenorhabditis elegans*. *Infection and Immunity, 73*, 5319–5328.

Quay, S. C., Friedman, S. B., & Eisenberg, R. C. (1972). Gluconate regulation of glucose catabolism in *Pseudomonas fluorescens*. *Journal of Bacteriology, 112*, 291–298.

Raaijmakers, J. M., de Bruijn, I., & de Kock, M. J. D. (2006). Cyclic lipopeptide production by plant-associated *Pseudomonas* spp.: Diversity, activity, biosynthesis, and regulation. *Molecular Plant-Microbe Interactions, 19*, 699–710.

Raaijmakers, J. M., Vlami, M., & de Souza, J. T. (2002). Antibiotic production by bacterial biocontrol agents. *Antonie van Leeuwenhoek, 81*, 537–547.

Rainey, P. B., & Bailey, M. J. (1996). Physical and genetic map of the *Pseudomonas fluorescens* SBW25 chromosome. *Molecular Microbiology, 19*, 521–533.

Ramette, A., Moënne-Loccoz, Y., & Défago, G. (2003). Prevalence of fluorescent pseudomonads producing antifungal phloroglucinols and/or hydrogen cyanide in soils naturally suppressive or conducive to tobacco black root rot. *FEMS Microbiology Ecology, 44*, 35–43.

Rasmussen, T. B., Skindersoe, M. E., Bjarnsholt, T., Phipps, R. K., Christensen, K. B., Jensen, P. O., Andersen, J. B., Koch, B., Larsen, T. O., Hentzer, M., Eberl, L., Høiby, N., & Givskov, M. (2005). Identity and effects of quorum-sensing inhibitors produced by *Penicillium* species. *Microbiology, 151*, 1325–1340.

Reimmann, C., Beyeler, M., Latifi, A., Winteler, H., Foglino, M., Lazdunski, A., & Haas, D. (1997). The global activator GacA of *Pseudomonas aeruginosa* PAO1 positively controls the production of the autoinducer *N*-butyryl-homoserine lactone and the formation of the virulence factors pyocyanin, cyanide, and lipase. *Molecular Microbiology, 24*, 309–319.

Reimmann, C., Valverde, C., Kay, E., & Haas, D. (2005). Posttranscriptional repression of GacS/GacA-controlled genes by the RNA-binding protein RsmE acting together with RsmA in the biocontrol strain *Pseudomonas fluorescens* CHA0. *Journal of Bacteriology, 187*, 276–285.

Rezzonico, F., Zala, M., Keel, C., Duffy, B., Moënne-Loccoz, Y. & Défago, G. (2007). Is the ability of biocontrol fluorescent pseudomonads to produce the antifungal metabolite 2,4-diacetylphloroglucinol really synonymous with higher plant protection? *New Phytologist, 173*, 861–872.

Riedel, K., Hentzer, M., Geisenberger, O., Huber, B., Steidle, A., Wu, H., Hoiby, N., Givskov, M., Molin, S., & Eberl, L. (2001). *N*-Acylhomoserine-lactone-mediated communication between *Pseudomonas aeruginosa* and *Burkholderia cepacia* in mixed biofilms. *Microbiology, 147*, 3249–3262.

Sambrook, J., & Russell, D. W. (2001). *Molecular cloning: A laboratory manual* (3rd ed.). Cold Spring Harbor, N.Y.: Cold Spring Harbor Laboratory Press.

Sarniguet, A., Kraus, J., Henkels, M. D., Muehlchen, A. M. & Loper, J. E. (1995). The sigma factor sigma s affects antibiotic production and biological control activity of *Pseudomonas fluorescens* Pf-5. *Proceedings of the National Academy of Sciences USA, 92*, 12255–12259.

Schmidli-Sacherer, P., Keel, C., & Défago, G. (1997). The global regulator GacA of *Pseudomonas fluorescens* CHA0 is required for suppression of root diseases in dicotyledons but not in *Gramineae*. *Plant Pathology, 46*, 80–90.

Schnider, U., Keel, C., Blumer, C., Troxler, J., Défago, G., & Haas, D. (1995). Amplification of the housekeeping sigma factor in *Pseudomonas fluorescens* CHA0 enhances antibiotic production and improves biocontrol abilities. *Journal of Bacteriology, 177*, 5387–5392.

Schnider-Keel, U., Seematter, A., Maurhofer, M., Blumer, C., Duffy, B., Gigot-Bonnefoy, C., Reimmann, C., Notz, R., Défago, G., Haas, D., & Keel, C. (2000). Autoinduction of 2,4-diacetylphloroglucinol biosynthesis in the biocontrol agent *Pseudomonas fluorescens* CHA0 and repression by the bacterial metabolites salicylate and pyoluteorin. *Journal of Bacteriology, 182*, 1215–1225.

Schouten, A., van den Berg, G., Edel-Hermann, V., Steinberg, C., Gautheron, N., Alabouvette, C., de Vos, C. H., Lemanceau, P., & Raaijmakers, J. M. (2004). Defense responses of *Fusarium oxysporum* to 2,4-diacetylphloroglucinol, a broad-spectrum antibiotic produced by *Pseudomonas fluorescens*. *Molecular Plant-Microbe Interactions, 17*, 1201–1211.

Sharifi-Tehrani, A., Zala, M., Natsch, A., Moënne-Loccoz, Y., & Défago, G. (1998). Biocontrol of soilborne fungal plant diseases by 2,4-diacetylphloroglucinol-producing fluorescent *Pseudomonas* with different restriction profiles of amplified 16S rDNA. *European Journal of Plant Pathology, 104*, 631–643.

Shaw, P. D., Ping, G., Daly, S. L., Cha, C., Cronan, J. E. Jr., Rinehart, K. L., & Farrand, S. K. (1997). Detecting and characterizing *N*-acyl-homoserine lactone signal molecules by thin-layer chromatography. *Proceedings of the National Academy of Sciences USA, 94*, 6036–6041.

Siddiqui, I. A., Haas, D., & Heeb, S. (2005). Extracellular protease of *Pseudomonas fluorescens* CHA0, a biocontrol factor with activity against the root-knot nematode *Meloidogyne incognita*. *Applied and Environmental Microbiology, 71*, 5646–5649.

Siqueira, J. O., Nair, M. G., Hammerschmidt, R., & Safir, G. R. (1991). Significance of phenolic compounds in plant-soil-microbial systems. *Critical Reviews in Plant Sciences, 10*, 63–121.

Somers, E., Vanderleyden, J., & Srinivasan, M. (2004). Rhizosphere bacterial signalling: A love parade beneath our feet. *Critical Reviews in Microbiology, 30*, 205–240.

Steidle, A., Sigl, K., Schuhegger, R., Ihring, A., Schmid, M., Gantner, S., Stoffels, M., Riedel, K., Givskov, M., Hartmann, A., Langebartels, C., & Eberl, L. (2001). Visualization of *N*-acylhomoserine lactone-mediated cell-cell communication between bacteria colonizing the tomato rhizosphere. *Applied and Environmental Microbiology, 67*, 5761–5770.

Stutz, E. W., Défago, G., & Kern, H. (1986). Naturally occurring fluorescent pseudomonads involved in suppression of black root rot of tobacco. *Phytopathology, 76*, 181–185.

Sutra, L., Siverio, F., Lopez, M. M., Hunault, G., Bollet, C., & Gardan, L. (1997). Taxonomy of *Pseudomonas* strains isolated from tomato pith necrosis: Amended description of *Pseudomonas corrugata* and proposal of three unnamed fluorescent *Pseudomonas* genomospecies. *International Journal of Systematic Bacteriology, 47*, 1020–1033.

Thomashow, L. S., & Weller, D. M. (1988). Role of a phenazine antibiotic from *Pseudomonas fluorescens* in biological control of *Gaeumannomyces graminis* var. *tritici*. *Journal of Bacteriology, 170*, 3499–3508.

Troxler, J., Azelvandre, P., Zala, M., Défago, G., & Haas, D. (1997). Conjugative transfer of chromosomal genes between fluorescent pseudomonads in the rhizosphere of wheat. *Applied and Environmental Microbiology, 63*, 213–219.

Turner, P., Barber, C., & Daniels, M. (1984). Behavior of the transposon Tn*5* and Tn*7* in *Xanthomonas campestris* pv. *campestris*. *Molecular and General Genetics, 195*, 101–107.

van den Broek, D., Chin-A-Woeng, T. F. C., Eijkemans, K., Mulders, I. H. M., Bloemberg, G. V., & Lugtenberg, B. J. J. (2003). Biocontrol traits of *Pseudomonas* spp. are regulated by phase variation. *Molecular Plant-Microbe Interactions, 16*, 1003–1012.

van Loon, L. C., Bakker, P. A. H. M., & Pieterse, C. M. J. (1998). Systemic resistance induced by rhizosphere bacteria. *Annual Review of Phytopathology, 36*, 453–483.

van Rij, E. T., Girard, G., Lugtenberg, B. J. J., & Bloemberg, G. V. (2005). Influence of fusaric acid on phenazine-1-carboxamide synthesis and gene expression of *Pseudomonas chlororaphis* strain PCL1391. *Microbiology, 151*, 2805–2814.

Ventre, I., Goodman, A. L., Vallet-Gely, I., Vasseur, P., Soscia, C., Molin, S., Bleves, S., Lazdunski, A., Lory, S., Filloux, A. (2006). Multiple sensors control reciprocal expression of *Pseudomonas aeruginosa* regulatory RNA and virulence genes. *Proceedings of the National Academy of Sciences USA, 103*, 171–176.

Wang, C., Zhang, H. B., Wang, L. H., & Zhang, L. H. (2006). Succinic semialdehyde couples stress response to quorum-sensing signal decay in *Agrobacterium tumefaciens*. *Molecular Microbiology, 62*, 45–56.

Wei, H. L., & Zhang, L. Q. (2006). Quorum-sensing system influences root colonization and biological control ability in *Pseudomonas fluorescens* 2P24. *Antonie van Leeuwenhoek, 267–280*, 1–14.

Weller, D. M. (1983). Colonization of wheat roots by a fluorescent pseudomonad suppressive to take-all. *Phytopathology, 73*, 1548–1553.

Weller, D. M., Landa, B. B., Mavrodi, O. V., Schroeder, K. L., De La Fuente, L., Blouin Bankhead, S., Allende Molar, R., Bonsall, R. F., Mavrodi, D. V., & Thomashow, L. S. (2007). Role of 2,4-diacetylphloroglucinol-producing fluorescent *Pseudomonas* spp. in the defense of plant roots. *Plant Biology, 9*, 4–20.

Weller, D. M., Raaijmakers, J. M., Gardener, B. B., & Thomashow, L. S. (2002). Microbial populations responsible for specific soil suppressiveness to plant pathogens. *Annual Review of Phytopathology, 40*, 309–348.

Whistler, C. A., & Pierson, L. S. 3rd. (2003). Repression of phenazine antibiotic production in *Pseudomonas aureofaciens* strain 30-84 by RpeA. *Journal of Bacteriology, 185*, 3718–3725.

Whistler, C. A., Stockwell, V. O., & Loper, J. E. (2000). Lon protease influences antibiotic production and UV tolerance of *Pseudomonas fluorescens* Pf-5. *Applied and Environmental Microbiology, 66*, 2718–2725.

Winzer, K., Hardie, K. R., & Williams, P. (2002). Bacterial cell-to-cell communication: Sorry, can't talk now—gone to lunch! *Current Opinion in Microbiology, 5*, 216–222.

Wood, D. W., Gong, F., Daykin, M., Williams, P., & Pierson, L. S. (1997). *N*-Acyl-homoserine lactone-mediated regulation of phenazine gene expression by *Pseudomonas aureofaciens* 30-84 in the wheat rhizosphere. *Journal of Bacteriology, 179*, 7663–7670.

Wood, D. W., & Pierson, L. S. (1996). The *phzI* gene of *Pseudomonas aureofaciens* 30-84 is responsible for the production of a diffusible signal required for phenazine antibiotic production. *Gene, 168*, 49–53.

Xavier, K. B., & Bassler, B. L. (2005). Interference with AI-2-mediated bacterial cell-cell communication. *Nature, 437*, 750–753.

Yahr, T. L., & Wolfgang, M. C. (2006). Transcriptional regulation of the *Pseudomonas aeruginosa* type III secretion system. *Molecular Microbiology, 62*, 631–640.

Zha, W., Rubin-Pitel, S. B., & Zhao, H. (2006). Characterization of the substrate specificity of PhlD, a type III polyketide synthase from *Pseudomonas fluorescens*. *Journal of Biological Chemistry, 281*, 32036–32047.

Zhang, L. H., & Dong, Y. H. (2004). Quorum sensing and signal interference: Diverse implications. *Molecular Microbiology, 53*, 1563–1571.

Zuber, S., Carruthers, F., Keel, C., Mattart, A., Blumer, C., Pessi, G., Gigot-Bonnefoy, C., Schnider-Keel, U., Heeb, S., Reimmann, C., & Haas, D. (2003). GacS sensor domains pertinent to the regulation of exoproduct formation and to the biocontrol potential of *Pseudomonas fluorescens* CHA0. *Molecular Plant-Microbe Interactions, 16*, 634–644.

Eur J Plant Pathol (2007) 119:329–339
DOI 10.1007/s10658-007-9162-4

REVIEW PAPER

Promotion of plant growth by ACC deaminase-producing soil bacteria

Bernard R. Glick · Zhenyu Cheng · Jennifer Czarny · Jin Duan

Received: 12 October 2006 / Accepted: 3 May 2007 / Published online: 5 July 2007

Abstract Plant growth-promoting bacteria that contain the enzyme 1-aminocyclopropane-1-carboxylate (ACC) deaminase facilitate plant growth and development by decreasing plant ethylene levels, especially following a variety of environmental stresses. In this review, the physiological basis for this growth-promotion effect is examined in some detail. In addition, models are presented that endeavour to explain (i) the seemingly paradoxical effects of ethylene on a plant's response to stress, (ii) how the expression of this enzyme is transcriptionally regulated in many bacterial strains and (iii) how ACC deaminase-containing plant growth-promoting bacteria alter plant gene expression and positively modulate plant growth.

Keywords ACC deaminase · Ethylene · Plant growth-promoting bacteria · Plant stress · Plant growth

Abbreviations

ACC	1-Aminocyclopropane-1-carboxylate
AOA	Aminooxyacetic acid
AVG	L-α-(aminoethoxyvinyl)-glycine
CRP	Cyclic AMP receptor protein
FNR	Fumarate–nitrate reduction regulatory protein
IAA	Indole-3-acetic acid
Lrp	Leucine-responsive regulatory protein
1-MCP	1-Methylcyclopropene
PAHs	Polycyclic aromatic hydrocarbons
PCBs	Polycyclic biphenyls
RAP PCR	RNA arbitrarily primed PCR

B. R. Glick (✉) · Z. Cheng · J. Czarny · J. Duan
Department of Biology, University of Waterloo,
200 University Avenue West, Waterloo, ON,
Canada N2L 3G1
e-mail: glick@uwaterloo.ca

Introduction

The growth of plants in the field may be inhibited by a large number of both biotic and abiotic stresses. These stresses include extremes of temperature, high light, flooding, drought, the presence of toxic metals and environmental organic contaminants, radiation, wounding, insect predation, high salt, and various pathogens including viruses, bacteria and fungi (Abeles et al. 1992). As a consequence of these environmental stresses, plant growth is invariably lower than it would be in their absence. Moreover, during its life, the plant is subject to a number of non-lethal stresses that limit plant growth until either the stress is removed or the plant is able to adjust its metabolism to overcome the stress so that, in the field, plant growth often consists of periods of maximal growth interspersed with periods of growth inhibition (Fig. 1).

In addition to the ability of a plant to modify its physiology and metabolism, including the synthesis

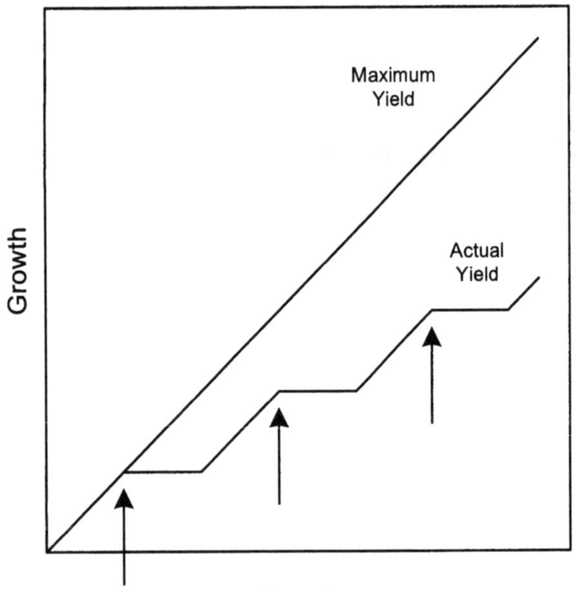

Fig. 1 Plant growth as a function of age. The slope of the maximum yield curve is the maximum growth rate. The arrows indicate the onset of a growth inhibitory, but non-lethal, stress which causes growth to cease or slow down for some period of time so that the actual yield is a direct result of the number and intensity of the stresses that a plant experiences during its lifetime

of a range of defensive proteins, certain soil bacteria can help plants to either avoid or partially overcome a variety of environmental stresses. These plant growth-promoting bacteria facilitate plant growth either by (i) aiding in the acquisition of nutritional resources such as nitrogen, phosphorus or iron; (ii) preventing the proliferation of pathogenic organisms (e.g. by synthesizing antibiotics); or by (iii) directly stimulating plant growth by either providing plant hormones such as auxin or cytokinin, or lowering plant ethylene levels through the action of the enzyme 1-aminocyclopropane-1-carboxylate (ACC) deaminase (Glick 1995; Glick et al. 1999).

Plant stress and ethylene

When plants are exposed to stressful conditions, they often respond by producing what is known as stress ethylene. Various plants respond differently to stress, however, and also have a range of sensitivities to ethylene. In addition, there is a complex web of interactions between ethylene and other plant hormones that varies somewhat from one plant to another, so that it is difficult to explain the functioning of stress ethylene in one simple model. Nevertheless, plants exposed to various types of stress invariably show increased ethylene levels leading, as a result to increased damage (Hyodo 1991).

In an apparent paradox, stress ethylene has been suggested to both alleviate and exacerbate some of the effects of pathogen infection, depending upon the plant species, its age and the nature of the pathogen (Abeles et al. 1992; Arshad and Frankenberger 2002; Van Loon and Glick 2004). A model that explains these seemingly contradictory effects of stress ethylene on plants has been proposed (Stearns and Glick 2003; Pierik et al. 2006; Van Loon et al. 2006). In one description of this model, there is an initial small peak of ethylene close in time, usually a few hours after, to the onset of the stress and then a second much larger peak some time later, usually one to three days (Fig. 2A). The first peak is only a small fraction of the magnitude of the second peak and is thought to initiate a protective response by the plant, such as transcription of pathogenesis-related genes and acquired resistance (Ciardi et al. 2000; Van Loon and Glick 2004). The first small wave of ethylene production is thought to consume the existing pool of ACC within plant tissues (Robison et al. 2001a). On the other hand, the second ethylene peak is so large that processes such as senescence, chlorosis and abscission are initiated, the overall effect of which is generally inhibitory to plant survival. Thus, following a severe infection by pathogens, a large portion of the damage that occurs to a plant is due to autocatalytic ethylene synthesis and not from direct pathogen action (Van Loon 1984). In this regard, not only can exogenous ethylene increase the severity of a pathogen infection, but as well, inhibitors of ethylene synthesis or ethylene action can significantly decrease the severity of a fungal or bacterial infection. The second peak of ethylene production occurs as a consequence of increased transcription of ACC synthase genes triggered by environmental and developmental cues (Yang and Hoffman 1984).

ACC deaminase

The enzyme ACC deaminase (EC: 4.1.99.4) which catalyzes the cleavage of ACC to ammonia and α-ketobutyrate was first discovered in 1978 (Honma and Shimomura 1978). This enzyme has subsequently

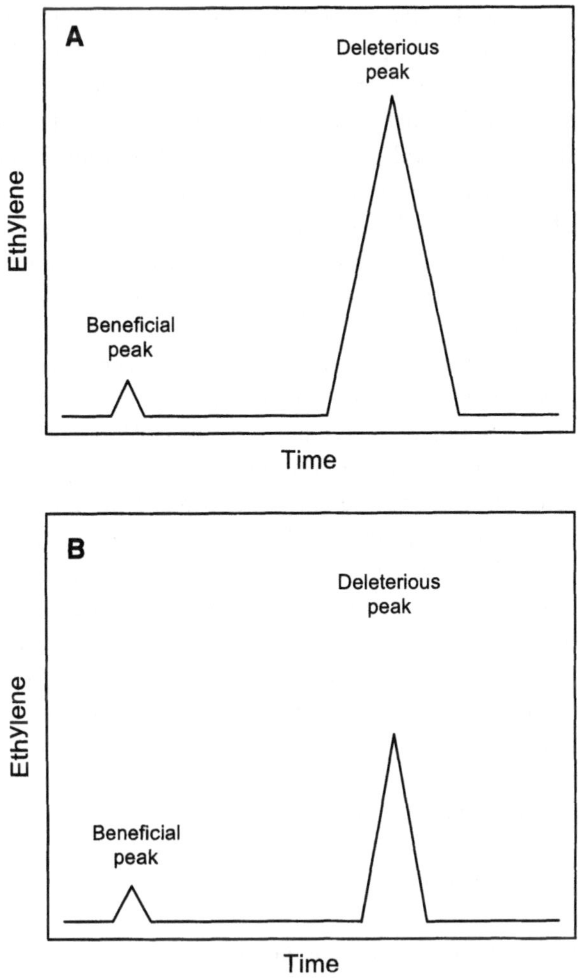

Fig. 2 Plant ethylene production as a function of time following an environmental stress. (**A**) In the absence of any exogenous bacteria. (**B**) In the presence of an ACC deaminase-producing plant growth-promoting bacterium. In both cases, there is an initial small peak of ethylene that is thought to activate transcription of plant defense genes, which is often difficult to detect, followed some time later by a much larger ethylene peak that can cause adverse responses in the plant. The amount of ethylene produced in response to an environmental stress is related to the plant age as well as the nature and severity of the stress

been detected in a wide range of bacterial strains and fungi (Klee et al. 1991; Sheehy et al. 1991; Honma 1993; Jacobson et al. 1994; Glick et al. 1995; Campbell and Thomson 1996; Burd et al. 1998; Minami et al. 1998; Jia et al. 1999; Belimov et al. 2001; Mayak et al. 2004a; Babalola et al. 2003; Ghosh et al. 2003; Ma et al. 2003a, b; Dey et al. 2004; Uchiumi et al. 2004; Belimov et al. 2005; Hontzeas et al. 2005; Blaha et al. 2006; Madhaiyan et al. 2006), and for many of these strains, the ACC deaminase gene has been isolated and characterized. The presence of ACC deaminase is relatively common amongst soil microorganisms. For example, in one study, 27 out of 233 newly isolated *Rhizobium* spp. from various sites in southern and central Saskatchewan contained this activity (Duan et al. 2006). In another study, ACC deaminase activity/genes were found in a wide range of bacterial isolates including *Azospirillum*, *Rhizobium*, *Agrobacterium*, *Achromobacter*, *Burkholderia*, *Ralstonia*, *Pseudomonas* and *Enterobacter* (Blaha et al. 2006). Moreover, 62 out of 88 *Pseudomonas* strains exhibiting biocontrol activity isolated from locations worldwide contained ACC deaminase (Wang et al. 2001).

K_m values of ACC deaminase for ACC have been estimated at pH 8.5, in all instances examined, to be approximately 1.5–17.4 mM indicating that the enzyme does not have a particularly high affinity for ACC (Honma and Shimomura 1978; Klee and Kishore 1992; Jacobson et al. 1994; Hontzeas et al. 2004a). Moreover, ACC levels in plants are typically in the μM range, therefore in most plant tissues the ACC concentration will be dramatically below the K_m of ACC deaminase for this substrate so that—based on the Michaelis-Menten rate equation for enzyme catalyzed reactions—a small increase in the ACC concentration (e.g. a doubling) will result in a parallel increase in the rate of ACC cleavage.

ACC deaminase is a member of a large group of enzymes that require the co-factor pyridoxal 5′-phosphate for enzymatic activity (Walsh et al. 1981). These enzymes have been classified based on their three dimensional structure, into four folding types: (i) tryptophan synthase, (ii) aspartate aminotransferase, (iii) D-amino acid aminotransferase and (iv) alanine racemase (Jansonius 1998). According to this classification scheme, ACC deaminase fits into the tryptophan synthase family. The coenzyme pyridoxal phosphate is a tightly bound cofactor of ACC deaminase in the amount of approximately one mole of pyridoxal phosphate per trimeric subunit (Honma 1985). Interestingly, ACC synthase also requires pyridoxal phosphate for enzyme activity.

ACC deaminase lowering of stress ethylene

A model was previously proposed by which plant growth-promoting bacteria can lower plant ethylene levels and in turn facilitate plant growth (Glick et al. 1998). In this model the plant growth-promoting bacteria bind to the surface of a plant (usually seeds or roots, although ACC deaminase-producing bacteria may also be found on leaves and flowers). In response to tryptophan and other small molecules in the plant exudates, the bacteria synthesize and secrete indole-3-acetic acid (IAA), some of which is taken up by the plant. This IAA together with endogenous plant IAA can stimulate plant cell proliferation, plant cell elongation or induce the transcription of ACC synthase which is the enzyme that catalyzes the formation of ACC. Some of the ACC is exuded from seeds, roots or leaves (Penrose et al. 2001; Grichko and Glick 2001a) along with other small molecules normally present in these exudates and may be taken up by the bacteria and subsequently cleaved by the enzyme, ACC deaminase, to ammonia and α-ketobutyrate. In this model, the bacterium acts as a sink for plant ACC and as a result of lowering either the endogenous or the IAA-stimulated ACC level, the amount of ethylene in the plant is also reduced. As a direct consequence of lowering plant ethylene levels, plant growth-promoting bacteria that possess the enzyme ACC deaminase can reduce the extent of ethylene inhibition of plant growth following a wide range of stresses. Thus, plants grown in association with these bacteria should have longer roots and shoots and be more resistant to growth inhibition by a variety of ethylene-inducing stresses.

The question arises, as to how bacterial ACC deaminase can selectively lower deleterious ethylene levels but not affect the small peak of ethylene that is thought to activate some plant defense responses (Fig. 2A). As discussed later in this review, ACC deaminase is generally present in bacteria at a low level until it is induced, and the induction of enzyme activity is a relatively slow and complex process. Immediately following an environmental stress, the pool of ACC in the plant is low as is the level of ACC deaminase in the associated bacterium. Following the relatively rapid induction of a low level of ACC oxidase in the plant, it is likely that there is increased flux through this enzyme resulting in the first small peak of ethylene which is of sufficient magnitude to induce a protective/defensive response in the plant (Fig. 2B). With time, bacterial ACC deaminase is induced (by the increasing amounts of ACC that ensue from the induction of ACC synthase in the plant) so that the magnitude of the second, deleterious, ethylene peak is decreased significantly (Fig. 2B). The second ethylene peak may be reduced dramatically, but it is never completely abolished since ACC oxidase has a much higher affinity for ACC than does ACC deaminase (Glick et al. 1998). Thus, when ACC deaminase-producing bacteria are present, ethylene levels are ultimately dependent upon the ratio of ACC oxidase to ACC deaminase (Glick et al. 1998).

Several different chemicals have been used to lower ethylene levels in plants including rhizobitoxine, L-α-(aminoethoxyvinyl)-glycine (AVG) which is a synthetic analog of rhiobitoxine, aminooxyacetic acid (AOA) and the ethylene perception inhibitor 1-methylcyclopropene (1-MCP) and its analogs (Yuhashi et al. 2000; Sisler and Serek 1997). While 1-MCP has been approved for commercial use, it has been utilized primarily to limit post-harvest fruit spoilage and flower wilting. However, unlike ACC deaminase-producing plant growth-promoting bacteria, chemical ethylene inhibitors cannot readily be used in the field as a means of limiting the inhibitory effects of biotic and abiotic stresses.

Phytopathogens

In recent years, the engineering of plants that are resistant to a variety of pathogens including viruses, bacteria and fungi has become popular. Unfortunately, it is impractical to attempt to engineer plants against all of the pathogens (and other stresses) that they might encounter in the environment, as these can vary from one locale to another and from one season to the next. Alternatively, one can either select or engineer biocontrol bacteria that protect plants against a range of different pathogens. A more general strategy, however, might include treating plant seeds or roots with plant growth-promoting or biocontrol bacteria that contain ACC deaminase. Thus, for example, ACC deaminase-producing biocontrol bacteria were more effective at preventing (i) growth inhibition of cucumber plants by the plant root pathogen *Pythium ultimum* and (ii) rotting of potatoes by *Erwinia carotovora* than were biocontrol bacteria that lacked this enzyme (Wang et al. 2000). In addition, transgenic tomato plants expressing a bacterial ACC deaminase gene under the transcriptional control of a root-specific promoter (which mimics the effect of adding ACC deaminase-producing plant growth-promoting bacteria to the plant roots) are significantly protected against damage from Verticillium wilt compared to non-transformed tomato plants (Robison et al. 2001b).

High salt and drought

Soil salinity is an enormous problem for agriculture under irrigation. In the hot and dry regions of the world the soils are frequently saline with low agricultural potential. In these areas most crops are grown under irrigation, and to exacerbate the problem, inadequate irrigation management leads to secondary salinization that affects 20% of irrigated land worldwide (Mayak et al. 2004b).

In recent years, considerable attention has been directed toward genetically engineering plants to be more salt tolerant, with moderate success (e.g. Apse et al. 1999). An alternative approach to overcoming some of the problems associated with growing plants in saline soils involves employing an ACC deaminase-producing bacterium, *Achromobacter piechaudii* ARV8, isolated from the rhizosphere of a *Lycium shawii* plant growing in a dry riverbed in the Arava region of Israel (Mayak et al. 2004b). This strain dramatically lowered the level of ethylene and prevented inhibition of plant growth in tomato plants grown in the presence of high concentrations of salt (Mayak et al. 2004b). The same bacterial strain lowered the ethylene level and significantly decreased the growth inhibition of peppers and tomatoes from drought stress (Mayak et al. 2004a).

Flooding

Periods of flooding can occur several times a growing season and may last for periods of from 1 or 2 days to several weeks. During these periods, the root environment rapidly becomes anaerobic causing an induction in the expression of ACC synthase, resulting in the accumulation of ACC in root tissues (Else and Jackson 1998). With other stresses, a significant portion of the newly synthesized ACC might be converted to ethylene in the roots; however, this is not possible when roots are flooded since the enzyme ACC oxidase, which catalyzes this reaction, requires oxygen for the conversion to proceed. Instead, the accumulated ACC is transported to the shoots where there is an aerobic environment and ethylene can be produced. Unfortunately, this causes epinasty, leaf chlorosis, necrosis and reduced fruit yield. On the other hand, when flooded plants are first treated with ACC deaminase-producing plant growth promoting bacteria, or plants are genetically engineered to express this enzyme in a root specific manner, much less ACC accumulates in the roots. Consequently, the damage to the plant that would otherwise occur from the newly synthesized ethylene is significantly decreased (Grichko and Glick 2001a, b).

Metal and organic contamination

In the presence of high levels of metals most plants synthesize growth inhibitory amounts of stress ethylene and also become severely iron depleted. This is readily remedied in the laboratory by adding ACC deaminase- and siderophore-producing plant growth-promoting bacteria which can help plants to overcome many of the effects of high levels of metal (Burd et al. 1998, 2000; Reed and Glick 2005). Similarly, transgenic plants that express a bacterial ACC deaminase gene under the control of a root-specific promoter are more resistant to the toxic effects of metals than are non-transformed plants (Grichko et al. 2000; Nie et al. 2002; Stearns et al. 2005; Li et al. 2006).

Field experiments aimed to facilitate plant growth in metal-contaminated soils so that the plant can take up and concentrate the metal (i.e., metal phytoremediation/phytoaccumulation) are considerably more complex than laboratory experiments. In the field, both ACC deaminase-producing plant growth-promoting bacteria and transgenic plants that express a bacterial ACC deaminase gene under the control of a root-specific promoter grow better than non-transformed and untreated plants. Although many metal contaminants are present at high levels in the field, in this environment they are generally not especially bioavailable so that only a small fraction of the metals are taken up by the plants (Farwell et al. 2006).

Considerable success has been achieved in the phytoremediation of organic environmental contaminants such as oil spills, polycyclic aromatic hydrocarbons (PAHs) and polycyclic biphenyls (PCBs) such that this technology is ready for commercialization. Many varieties of plants and trees can take up and degrade some organic compounds; however, larger molecules are less water soluble and more difficult to degrade, often requiring degradative bacteria as well as plant roots for their breakdown. While the degradative bacterial population in the bulk soil is insufficient to efficiently break down complex organic molecules, the bacterial population in the rhizosphere is typically 100–1,000 times greater than in bulk soil so that most of the degradation of organic environmental contaminants occurs in the rhizosphere. Despite the fact that many plants, together with rhizosphere degradative bacteria, can readily degrade many organic environmental pollutants, most of these compounds are somewhat inhibitory to plant growth. Not surprisingly, a significant part of this growth inhibition is a consequence of the production of stress ethylene by the plant. Thus, treatment of plant seeds or roots with ACC deaminase-producing plant growth-promoting bacteria relieves much of this growth inhibition, allowing the plant to grow to near normal size, and degradation of the contaminants to proceed at a much faster rate than would otherwise be possible (Huang et al. 2004; Reed and Glick 2005; Huang et al. 2005; Greenberg et al. 2006).

Rhizobial infection

Ethylene is an inhibitor of rhizobial nodulation of legumes, and since the infection of plant roots by *Rhizobia* causes plants to locally produce ethylene, rhizobial infection may be viewed as a self-limiting process (Guinel and Geil 2002; Ma et al. 2003a). On the other hand, many strains of *Rhizobia* produce either rhizobitoxine, an inhibitor of the enzyme ACC synthase, or ACC deaminase which allows these bacterial strains to lower the ethylene levels and increase nodulation (and subsequent biomass formation) by 25–40% (Nukui et al. 2000; Ma et al. 2003b, 2004). Moreover, surveying *Rhizobia* strains for ACC deaminase activity indicates that a large number of commercial strains but only a small number of field strains have this activity. This suggests that the direct screening of field isolates for ACC deaminase activity may be one means of rapidly selecting *Rhizobia* strains with superior commercial inoculant potential.

Rhizobia strains that express ACC deaminase exhibit only a low level of enzyme activity compared with free-living plant growth-promoting bacteria. This has led us to speculate that there are two types of ACC deaminase-producing bacteria. On the one hand, there are free-living bacteria that bind relatively non-specifically to plant tissues (mainly roots) and have a high level of ACC deaminase activity which protects plants from a range of different stresses by lowering ethylene levels throughout the plant. On the other hand, *Rhizobia* bind tightly only to the roots of specific plants and have a low level of enzyme activity which facilitates nodulation by locally (but not globally) lowering ethylene levels. At this point, it is not known whether the 10- to 30-fold differences in enzyme activity observed when comparing free-living bacteria with *Rhizobia* is a consequence of differences in the amount of enzyme synthesized or in the intrinsic catalytic activity of the enzymes from different types of bacteria.

ACC deaminase regulation

Full induction of ACC deaminase gene expression requires the addition of ACC to the growing cells and takes much longer than the generation time of the bacterium (Jacobson et al. 1994). This suggests that ACC deaminase induction, and hence its mode of regulation in this bacterium, is relatively complex. In fact, analysis of DNA sequence data for the region upstream of the ACC deaminase structural gene (*acdS*) from *Pseudomonas putida* UW4 indicates that this DNA segment contains a CRP (cyclic AMP receptor protein) binding site, an FNR (fumarate–nitrate reduction regulatory protein) binding site (a known anaerobic transcriptional regulator), an Lrp (leucine-responsive regulatory protein) binding site, an open reading frame encoding an Lrp protein and three putative promoter sequences, one controlling the ACC deaminase regulatory gene (*acdR*; encoding Lrp) and two controlling *acdS* (Grichko and Glick 2000; Li and Glick 2001). All of these features were shown to be involved in the transcriptional regulation of *acdS*. More recently, in this same bacterium, a protein (AcdB) that interacts directly with ACC, the Lrp protein and the region of DNA upstream of *acdS* was identified and characterized (Z. Cheng, B.P. Duncker, B. McConkey and B.R. Glick submitted for publication). Although all of the details of how these various proteins and regions of DNA interact have not been completely elaborated, based on a combination of the published and submitted data, a model of the transcriptional regulation of *acdS* has been developed (Fig. 3). In addition to *P. putida* UW4, genes encoding Lrp proteins have been found immediately upstream from a number of bacterial ACC deaminase structural genes, and in every instance *acdS* and *acdR* were oriented in opposite directions (N. Hontzeas, J. Duan and B.R. Glick unpublished results). In many instances, however, neither the CRP nor the FNR binding site were found to be present. These data suggest that *acdS* and *acdR* are usually inherited together and that this mode of transcriptional regulation is a central feature of the functioning of many bacterial ACC deaminases.

In addition to the more common mode of regulation by AcdR (= Lrp), the *acdS* gene from *Mesorhizobium loti* MAFF303099 was found to be under the control of a *nifA* promoter (Uchiumi et al. 2004) and to be expressed within legume nodules (Nukui et al. 2006). This observation leads to the speculation that the expression of *acdS* within nitrogen-fixing nodules might act to decrease the rate at which the nodule senesces—as a consequence of its high energy demand, nitrogen fixation could activate stress ethylene synthesis—and the resultant longer nodule lifetime might effectively increase the amount of fixed nitrogen.

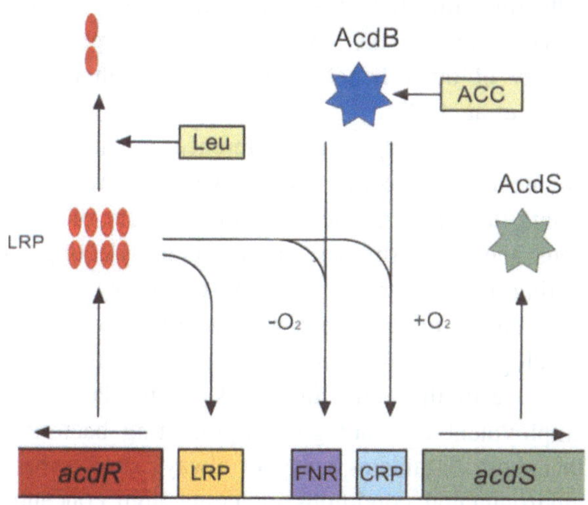

Fig. 3 Model of the transcriptional regulation of ACC deaminase expression in *Pseudomonas putida* UW4. The *acdR* gene encodes an Lrp protein which is thought to function as an octamer (Leonard et al. 2001). This protein can either bind to a DNA sequence known as an LRP box (which overlaps the promoter for *acdR*, not shown), preventing further transcription of this gene, or it can bind to a complex of ACC and the AcdB protein, encoding glycerophosphoryl diester phosphodiesterase (Z. Cheng, B.P. Duncker, B. McConkey and B.R. Glick submitted for publication), and together Lrp and AcdB can bind to either an FNR or CRP box on the DNA (both of which overlap separate promoter sequences, not shown). Binding to FNR is favoured under anerobic conditions while binding to CRP is favoured under aerobic conditions. The binding of these factors facilitates transcription of *acdS* by RNA polymerase. The newly synthesized ACC deaminase (= AcdS) cleaves ACC to form ammonia and α-ketobutyrate with the latter compound being a precursor of branched chain amino acids including leucine. Finally, in the presence of high levels of leucine in the cell, the Lrp octamer is dissociated into an inactive dimeric form thereby shutting down further transcription of *acdS*

Effect of ACC deaminase on plant gene expression

In one study, differential display PCR was used to elaborate some of the changes in plant gene expression caused by the addition of the plant growth-promoting bacterium *Paenibacillus polymyxa* to the roots of *Arabidopsis thaliana* plants (Timmusk and Wagner 1999). These workers identified a small number of genes whose expression was altered significantly and concluded that the plant responded to the presence of the bacterium as if the bacterium was a mild biotic stress.

In a subsequent study, RNA arbitrarily primed (RAP) PCR was used to identify several genes in canola roots whose expression was affected differentially by the addition of an ACC deaminase-producing plant growth-promoting bacterium and an ACC deaminase negative mutant of that strain (Hontzeas et al. 2004b). Interestingly, the ACC deaminase-producing bacterium down-regulated genes involved in ethylene-induced plant stress responses in the plant and up-regulated genes involved in plant growth. These data are consistent with the notion, when plant growth-promoting bacteria express ACC deaminase they are no longer perceived by the plant as a mild biotic stress.

Recently, canola shoot mRNA was isolated and hybridized to microarrays in which each chip contained more than 20,000 different 60-mer oligonucleotide DNA sequences representing approximately 80% of the *Arabidopsis* genome. In these experiments, non-transformed canola was compared to transgenic canola expressing a bacterial ACC deaminase gene under the control of a root-specific promoter. Analysis of the results indicated that several auxin response factor genes were more highly transcribed in the transgenic plants (J. Czarny, S. Shah and B.R. Glick unpublished results). While these preliminary experiments need to be repeated, and gene expression needs to be examined in canola plants treated with ACC deaminase-producing bacteria, it is intriguing to speculate as to how these data fit into our understanding of the mode of action of ACC deaminase. One model (Fig. 4) that explains the data is that by lowering the ethylene concentration in plant roots, ACC deaminase relieves the ethylene repression of auxin response factor synthesis, and indirectly increases plant growth (see also Dharmasiri and Estelle 2004). This interaction between ethylene and IAA acts as a feedback loop which decreases the impact of IAA and also may decrease the amount of IAA-stimulated ethylene synthesis that might otherwise occur. In this way, ethylene may limit the amount of its own synthesis. In concert with the model proposed here, it has been reported that ethylene can inhibit the transport of IAA in various plants (Burg and Burg 1966; Morgan and Gausman 1966; Suttle 1988; Prayitno et al. 2006).

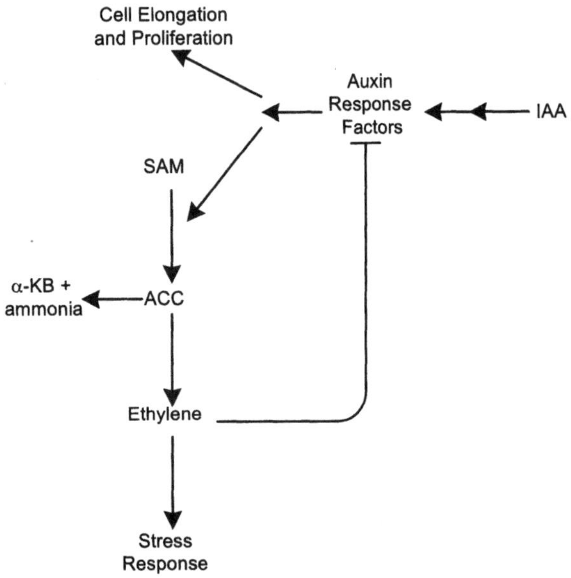

Fig. 4 Model for how the ACC deaminase lowering of ethylene levels leads to an increase in IAA-mediated canola gene expression. While this model was developed from microarray data comparing transgenic plants expressing ACC deaminase under the control of a root-specific promoter, based on physiological data, it is expected that non-transformed plants treated with an ACC deaminase-producing plant growth-promoting bacterium will behave similarly. Here, in the absence of ACC deaminase, root-produced ethylene inhibits transcription of auxin response factors thereby limiting auxin-stimulated plant growth as well as auxin promotion of ACC synthase transcription. In the presence of ACC deaminase, ethylene levels are decreased and the blockage of auxin response factor transcription is relieved thereby facilitating plant growth

Conclusions and future prospects

All of the available data are consistent with the previously proposed model of plant growth facilitation by ACC deaminase-producing plant growth-promoting bacteria (Glick et al. 1998). Moreover, plants respond similarly to ACC deaminase regardless of whether the enzyme is expressed in the roots of transgenic plants or as part of a root-associated bacterium. If anything, the root-associated bacterium provides a greater benefit to the plant, most likely reflecting the fact that in addition to lowering ethylene levels, the bacteria may also provide a variety of other benefits to the plant (Glick 1995; Glick et al. 1999).

Given the very large difference in the cost of engineering, selecting and developing transgenic plants that are protected against a variety of pathogens and other stresses, compared to selecting and testing appropriate plant growth-promoting bacteria, it is more propitious to direct our efforts toward the development of new plant growth-promoting bacteria. One of the major drawbacks in the large scale employment of plant growth-promoting bacteria is that these organisms may not always survive harsh environmental conditions including high concentrations of environmental contaminants, salts, extremes of pH and temperature, and the presence of other organisms that either out-compete or consume these bacteria. A possible solution to this problem may lie in the use of endophytic plant growth-promoting bacteria (Sturz and Nowak 2000).

Acknowledgements The work described in this review was funded by the Natural Sciences and Engineering Research Council of Canada, to B.R.G. and a fellowship to J.C. We thank Dr. Elisa Gamalero for critically reading the manuscript.

References

Abeles, F. B., Morgan, P. W., & Saltveit, M. E. Jr. (1992). *Ethylene in plant biology*. New York: Academic Press.
Apse, M. P., Aharon, G. S., Snedden, W. A., & Blumwald, E. (1999). Salt tolerance conferred by overexpression of a

vacuolar Na+/H+ antiport in *Arabidopsis*. *Science, 285*, 1256–1258.

Arshad, M., & Frankenberger, W. T. Jr. (2002). *Ethylene: Agricultural sources and applications*. Dordrecht, The Netherlands: Kluwer Academic/Plenum Publishers.

Babalola, O. O., Osir, E. O., Sanni, A. I., Odhaimbo, G. D., & Bulimo, W. D. (2003). Amplification of 1-aminocyclopropane-1-carboxylic (ACC) deaminase from plant growth promoting rhizobacteria in Striga-infested soils. *African Journal of Biotechnology, 2*, 157–160.

Belimov, A. A., Hontzeas, N., Safronova, V. I., Demchinskaya, S. V., Piluzza, G., Bullitta, S., & Glick, B. R. (2005). Cadmium-tolerant plant growth-promoting rhizobacteria associated with the roots of Indian mustard (*Brassica juncea* L. Czern.). *Soil Biology and Biochemistry, 37*, 241–250.

Belimov, A. A., Safronova, V. I., Sergeyeva, T. A., Egorova, T. N., Matveyeva, V. A., Tsyganov, V. E., Borisov, A. Y., Tikhonovich, I. A., Kluge, C., Preisfeld, A., Dietz, K. J., & Stepanok, V. V. (2001). Characterization of plant growth promoting rhizobacteria isolated from polluted soils and containing 1-aminocyclopropane-1-carboxylate deaminase. *Canadian Journal of Microbiology, 47*, 642–652.

Blaha, D., Prigent-Combaret, C., Mirza, M. S., & Moënne-Loccoz, Y. (2006). Phylogeny of the 1-aminocyclopropane-1-carboxylic acid deaminase-encoding gene *acdS* in phytobeneficial and pathogenic *Proteobacteria* and relation with strain biogeography. *FEMS Microbiology Ecology, 56*, 455–470.

Burd, G. I., Dixon, D. G., & Glick, B. R. (1998). A plant growth-promoting bacterium that decreases nickel toxicity in seedlings. *Applied and Environmental Microbiology, 64*, 3663–3668.

Burd, G. I., Dixon, D. G., & Glick, B. R. (2000). Plant growth-promoting bacteria that decrease heavy metal toxicity in plants. *Canadian Journal of Microbiology, 46*, 237–245.

Burg, S. P., & Burg, E. A. (1966). The interaction between auxin and ethylene and its role in plant growth. *American Journal of Botany, 55*, 262–269.

Campbell, B. G., & Thomson, J. A. (1996). 1-Aminocyclopropane-1-carboxylate deaminase genes from *Pseudomonas* strains. *FEMS Microbiology Letters, 138*, 207-210.

Ciardi, J. A., Tieman, D. M., Lund, S. T., Jones, J. B., Stall, R. E., & Klee, H. J. (2000). Response to *Xanthomonas campestris* pv. *vesicatoria* in tomato involves regulation of ethylene receptor gene expression. *Plant Physiology, 123*, 81–92.

Dey, R., Pal, K. K., Bhatt, D. M., & Chauhan, S. M. (2004). Growth promotion and yield enhancement of peanut (*Aracis hypoggaea* L.) by application of plant growth-promoting rhizobacteria. *Microbiological Research, 159*, 371–394.

Dharmasiri, N., & Estell, M. (2004). Auxin signaling and regulated protein degradation. *Trends in Plant Science, 9*, 302–308.

Duan, J., Müller, K. M., Charles, T. C., Vesely, S., & Glick, B. R. (2006). 1-Aminocyclopropane-1-carboxylate (ACC) deaminase genes in *Rhizobia*: Isolation, characterization and regulation. Proceedings of the 7th International PGPR Workshop (50 pp). Amsterdam.

Else, M. A., & Jackson, M. B. (1998). Transport of 1-aminocyclopropane-1-carboxylic acid (ACC) in the transpiration stream of tomato (*Lycopersicon esculentum*) in relation to foliar ethylene production and petiole epinasty. *Australian Journal of Plant Physiology, 25*, 453–458.

Farwell, A. J., Vesely, S., Nero, V., Rodriguez, H., Shah, S., Dixon, D. G., & Glick, B. R. (2006). The use of transgenic canola (*Brassica napus*) and plant growth-promoting bacteria to enhance plant biomass at a nickel-contaminated field site. *Plant and Soil, 288*, 309–318.

Ghosh, S., Penterman, J. N., Little, R. D., Chavez, R., & Glick, B. R. (2003). Three newly isolated plant growth-promoting bacilli facilitate the growth of canola seedlings. *Plant Physiology and Biochemistry, 41*, 277–281.

Glick, B. R. (1995). The enhancement of plant growth by free-living bacteria. *Canadian Journal of Microbiology, 41*, 109–117.

Glick, B. R., Karaturovíc, D. M., & Newell, P. C. (1995). A novel procedure for rapid isolation of plant growth promoting pseudomonads. *Canadian Journal of Microbiology, 41*, 533–536.

Glick, B. R., Patten, C. L., Holguin, G., & Penrose, D. M. (1999). *Biochemical and genetic mechanisms used by plant growth promoting bacteria*. London: Imperial College Press.

Glick, B. R., Penrose, D. M., & Li, J. (1998). A model for the lowering of plant ethylene concentrations by plant growth-promoting bacteria. *Journal of Theoretical Biology, 190*, 63–68.

Greenberg, B. M., Huang, X. D., Gurska, Y., Gerhardt, K. E., Wang, W., Lampi, M. A., Zhang, C., Khalid, A., Isherwood, D., Chang, P., Wang, H., Dixon, D. G., & Glick, B. R. (2006). Successful field tests of a multi-process phytoremediation system for decontamination of persistent petroleum and organic contaminants, Proceedings of the 29th Arctic and Marine Oil Spill Program Technical Seminar (Vol. 1, pp. 389–400).

Grichko, V. P., Filby, B., & Glick, B. R. (2000). Increased ability of transgenic plants expressing the bacterial enzyme ACC deaminase to accumulate Cd, Co, Cu, Ni, Pb and Zn. *Journal of Biotechnology, 81*, 45-53.

Grichko, V. P., & Glick, B. R. (2000). Identification of DNA sequences that regulate the expression of the *Enterobacter cloacae* UW4 1-aminocyclopropane-1-carboxylate deaminase gene. *Canadian Journal of Microbiology, 46*, 1159–1165.

Grichko, V. P., & Glick, B. R. (2001a). Amelioration of flooding stress by ACC deaminase-containing plant growth-promoting bacteria. *Plant Physiology and Biochemistry, 39*, 11–17.

Grichko, V. P., & Glick, B. R. (2001b). Flooding tolerance of transgenic tomato plants expressing the bacterial enzyme ACC deaminase controlled by the 35*S*, *rolD* or PRB-1*b* promoter. *Plant Physiology and Biochemistry, 39*, 19–25.

Guinel, F. C., & Geil, R. D. (2002). A model for the development of the rhizobial and arbuscular mycorrhizal symbioses in legumes and its use to understand the roles of ethylene in the establishment of these two symbioses. *Canadian Journal of Botany, 80*, 695–720.

Honma, M. (1985). Chemically reactive sulfhydryl groups of 1-aminocyclopropane-1-carboxylate deaminase. *Agricultural and Biological Chemistry, 49*, 567–571.

Honma, M. (1993). Stereospecific reaction of 1-aminocyclopropane-1-carboxylate deaminase. In J. C. Pech, A. Latché, & C. Balagué (Eds.), *Cellular and molecular aspects of the plant hormone ethylene* (pp. 111–116). Dordrecht, The Netherlands: Kluwer Academic Publishers.

Honma, M., & Shimomura, T. (1978). Metabolism of 1-aminocyclopropane-1-carboxylic acid. *Agricultural and Biological Chemistry, 42*, 1825–1831.

Hontzeas, N., Richardson, A. O., Belimov, A. A., Safranova, V. I., Abu-Omar, M. M., & Glick, B. R. (2005). Evidence for horizontal gene transfer (HGT) of ACC deaminase genes. *Applied and Environmental Microbioogy, 71*, 7556–7558.

Hontzeas, N., Zoidakis, J., Glick, B. R., & Abu-Omar, M. M. (2004a). Expression and characterization of 1-aminocyclopropane-1-carboxylate deaminase from the rhizobacterium *Pseudomonas putida* UW4: A key enzyme in bacterial plant growth promotion. *Biochimica et Biophysica Acta, 1703*, 11–19.

Hontzeas, N., Saleh, S. S., & Glick, B. R. (2004b). Changes in gene expression in canola roots induced by ACC deaminase-containing plant growth-promoting bacteria. *Molecular Plant–Microbe Interactions, 17*, 865–871.

Huang, X.-D., El-Alawai, Y., Gurska, J., Glick, B. R., & Greenberg, B. M. (2005). A multi-process phytoremediation system for decontamination of persistent total petroleum hydrocarbons (TPHs) from soils. *Microchemical Journal, 81*, 139–147.

Huang, X.-D., El-Alawi, Y., Penrose, D. M., Glick, B. R., & Greenberg, B. M. (2004). Responses of plants to creosote during phytoremediation and their significance for remediation processes. *Environmental Pollution, 130*, 453–463.

Hyodo, H. (1991). Stress/wound ethylene. In A. K. Mattoo, & J. C. Shuttle (Eds.), *The plant hormone ethylene* (pp. 65–80). Boca Raton: CRC Press.

Jacobson, C. B., Pasternak, J. J., & Glick, B. R. (1994). Partial purification and characterization of 1-aminocyclopropane-1-carboxylate deaminase from the plant growth promoting rhizobacterium *Pseudomonas putida* GR12-2. *Canadian Journal of Microbiology, 40*, 1019–1025.

Jansonius, N. J. (1998). Structure, evolution and action of vitamin B6-dependent enzymes. *Current Opinion in Structural Biology, 8*, 759–769.

Jia, Y. J., Kakuta, Y., Sugawara, M., Igarashi, T., Oki, N., Kisaki, M., Shoji, T., Kanetuna, Y., Horita, T., Matsui, H., & Honma, M. (1999). Synthesis and degradation of 1-aminocyclopropane-1-carboxylic acid by *Penicillium citrinum. Bioscience, Biotechnology and Biochemistry, 63*, 542–549.

Klee, H. J., Hayford, M. B., Kretzmer, K. A., Barry, G. F., & Kishore, G. M. (1991). Control of ethylene synthesis by expression of a bacterial enzyme in transgenic tomato plants. *Plant Cell, 3*, 1187–1193.

Klee, H. J., & Kishore, G. M. (1992). Control of fruit ripening and senescence in plants. United States Patent Number: 5,702,933.

Leonard, P. M., Smits, S. H. J., Sedelnikova, S. E., Brinkman, A. B., de Vos, W. M., van der Oost, J., Rice, D. W., & Rafferty, J. B. (2001). Crystal structure of the Lrp-like transcrptional regulator from the archaeon *Pyrococcus furiosus. EMBO Journal, 20*, 990–997.

Li, J., & Glick, B. R. (2001). Transcriptional regulation of the *Enterobacter cloacae* UW4 1-aminocyclopropane-1-carboxylate (ACC) deaminase gene (*acdS*). *Canadian Journal of Microbiology, 47*, 359–367.

Li, Q., Shah, S., Saleh-Lakha, S., & Glick, B. R. (2006). Growth of tobacco in nickel-contaminated soil in the presence of the plant growth-promoting bacterium *Pseudomonas putida* UW4. Current Microbiology (in press).

Ma, W., Charles, T. C., & Glick, B. R. (2004). Expression of an exogenous 1-aminocyclopropane-1-carboxylate deaminase gene in *Sinorhizobium meliloti* increases its ability to nodulate alfalfa. *Applied and Environmental Microbiology, 70*, 5891–5897.

Ma, W., Guinel, F. C., & Glick, B. R. (2003b). The *Rhizobium leguminosarum* bv. *viciae* ACC deaminase protein promotes the nodulation of pea plants. *Applied and Environmental Microbiology, 69*, 4396–4402.

Ma, W., Sebestianova, S., Sebestian, J., Burd, G. I., Guinel, F., & Glick, B. R. (2003a). Prevalence of 1-aminocyclopropaqne-1-carboxylate in deaminase in *Rhizobia* spp. *Antonie Van Leeuwenhoek, 83*, 285–291.

Madhaiyan, M., Poonguzhali, S., Ryu, J., & Sa, T. (2006). Regulation of ethylene levels in canola (*Brassica campestris*) by 1-aminocycloprpane-1-carboxylate deaminase-containing *Methylobacterium fjisawaense. Planta, 224*, 268–278.

Mayak, S., Tirosh, T., & Glick, B. R. (2004a). Plant growth-promoting bacteria that confer resistance to water stress in tomato and pepper. *Plant Science, 166*, 525–530.

Mayak, S., Tirosh, T., & Glick, B. R. (2004b). Plant growth-promoting bacteria that confer resistance in tomato to salt stress. *Plant Physiology and Biochemistry, 42*, 565–572.

Minami, R., Uchiyama, K., Murakami, T., Kawai, J., Mikami, K., Yamada, T., Yokoi, D., Ito, H., Matsui, H., & Honma, M. (1998). Properties, sequence, and synthesis in *Escherichia coli* of 1-aminocyclopropane-1-carboxylate deaminase from *Hansenula saturnus. Journal of Biochemistry, 123*, 1112–1118.

Morgan, P. W., & Gausman, H. W. (1966). Effects of ethylene on auxin transport. *Plant Physiology, 41*, 45–52.

Nie, L., Shah, S., Burd, G. I., Dixon, D. G., & Glick, B. R. (2002). Phytoremediation of arsenate contaminated soil by transgenic canola and the plant growth-promoting bacterium *Enterobacter cloacae* CAL2. *Plant Physiology and Biochemistry, 40*, 355–361.

Nukui, N., Ezura, H., Yuhashi, K., Yasuta, T., & Minamisawa, K. (2000). Effects of ethylene precursor and inhibitors for ethylene biosynthesis and perception on nodulation in *Lotus japonicus* and *Macroptilium atropurpureum. Plant Cell Physiology, 41*, 893–897.

Nukui, N., Minamisawa, K., Ayabe, S. I., & Aoki, T. (2006). Expression of the 1-aminocyclopropane-1-carboxylic acid deaminase gene requires symbiotic nitrogen-fixing regulator gene *nifA2* in *Mesorhizobium loti* MAFF303099. *Applied and Environmental Microbiology, 72*, 4964–4969.

Penrose, D. M., Moffatt, B. A., & Glick, B. R. (2001). Determination of 1-aminocyclopropane-1-carboxylic acid (ACC) to assess the effects of ACC deaminase-containing bacteria on roots of canola seedlings. *Canadian Journal of Microbiology, 47*, 77–80.

Pierik, R., Tholen, D., Poorter, H., Visser, E. J. W., & Voesenek, L. A. C. J. (2006). The Janus face of ethylene: Growth inhibition and stimulation. *Trends in Plant Science, 11*, 176–183.

Prayitno, J., Rolfe, B. G., & Mathesius, U. (2006). The ethylene-insensitive *sickle* mutant of *Medicago truncatula* shows altered auxin transport regulation during nodulation. *Plant Physiology, 142*, 168–180.

Reed, M. L. E., & Glick, B. R. (2005). Growth of canola (*Brassica napus*) in the presence of plant growth-promoting bacteria and either copper or polycyclic aromatic hydrocarbons. *Canadian Journal of Microbiology, 51*, 1061–1069.

Robison, M. M., Griffith, M., Pauls, K. P., & Glick, B. R. (2001a). Dual role of ethylene in susceptibility of tomato to Verticillium wilt. *Journal of Phytopathology, 149*, 385–388.

Robison, M. M., Shah, S., Tamot, B., Pauls, K. P., Moffatt, B. A., & Glick, B. R. (2001b). Reduced symptoms of Verticillium wilt in transgenic tomato expressing a bacterial ACC deaminase. *Molecular Plant Pathology, 2*, 135–145.

Sheehy, R. E., Honma, M., Yamada, M., Sasaki, T., Martineau, B., & Hiatt, W. R. (1991). Isolation, sequence, and expression in *Escherichia coli* of the *Pseudomonas* sp. strain ACP gene encoding 1-aminocyclopropane-1-carboxylate deaminase. *Journal of Bacteriology, 173*, 5260–5265.

Sisler, E. C., & Serek, M. (1997). Inhibitors of ethylene responses in plants at the receptor level: Recent developments. *Physiologia Plantarum, 100*, 577–582.

Stearns, J., & Glick, B. R. (2003). Transgenic plants with altered ethylene biosynthesis or perception. *Biotechnology Advances, 21*, 193–210.

Stearns, J. C., Shah, S., Dixon, D. G., Greenberg, B. M., & Glick, B. R. (2005). Tolerance of transgenic canola expressing 1-aminocyclopropane-carboxylic acid deaminase to growth inhibition by nickel. *Plant Physiology and Biochemistry, 43*, 701–708.

Sturz, A. V., & Nowak, J. (2000). Endophytic communities of rhizobacteria and the strategies required to create yield enhancing associations with crops. *Applied Soil Ecology, 15*, 183–190.

Suttle, J. C. (1988). Effect of IAA on polar IAA transport, net IAA uptake and specific binding of *N*-1-naphthylphthalkamic in tissues and microsomes isolated from etiolated pea epicotyls. *Plant Physiology, 88*, 795–799.

Timmusk, S., & Wagner, E. G. H. (1999). The plant growth promoting rhizobacterium *Paenibacillus polymyxa* induces changes in *Arabidopsis thaliana* gene expression: A possible connection between biotic and abiotic stress responses. *Molecular Plant–Microbe Interactions, 12*, 951–959.

Uchiumi, T., Ohwada, T., Itakura, M., Mitsui, H., Nukui, N., Dawadi, P., Kaneko, T., Tabata, S., Yokoyama, T., Tejima, K., Saeki, K., Omori, H., Hayashi, M., Maekawa, T., Sriprang, R., Murooka, Y., Tajima, S., Simomura, K., Nomura, M., Suzuki, A., Shimoda, Y., Sioya, K., Abe, M., & Minamisawa, K. (2004). Expression islands clustered on the symbiosis island of the *Mesorhizobium loti* genome. *Journal of Bacteriology, 186*, 2439–2448.

Van Loon, L. C. (1984). Regulation of pathogenesis and symptom expression in diseased plants by ethylene. In Y. Fuchs, & E. Chalutz (Eds.), *Ethylene: Biochemical, physiological and applied aspects* (pp. 171–180). The Hague: Martinus Nijhoff/Dr W. Junk.

Van Loon, L. C., Geraats, B. P. J., & Linthorst, H. J. M. (2006). Ethylene as a modulator of disease resistance in plants. *Trends in Plant Science, 11*, 184–191.

Van Loon, L. C., & Glick, B. R. (2004). Increased plant fitness by rhizobacteria. In H. Sandermann (Ed.), *Molecular ecotoxicology of plants* (pp. 177–205). Berlin: Springer-Verlag.

Walsh, C., Pascal, R. A., Johnston, M., Raines, R., Dikshit, D., Krantz, A., & Honma, M. (1981). Mechanistic studies on the pyridoxal phosphate enzyme 1-aminocyclopropane-1-carboxylate from *Pseudomonas* sp. *Biochemistry, 20*, 7509–7519.

Wang, C., Knill, E., Glick, B. R., & Défago, G. (2000). Effect of transferring 1-aminocyclopropane-1-carboxylic acid (ACC) deaminase genes into *Pseudomonas fluorescens* strain CHA0 and its *gacA* derivative CHA96 on their growth promoting and disease-suppressive capacities. *Canadian Journal of Microbiology, 46*, 898–907.

Wang, C., Ramette, A., Punjasamarnwong, P., Zala, M., Natsch, A., Moënne-Loccoz, Y., & Défago, G. (2001). Cosmopolitan distribution of *phlD*-containing dicotyledonous crop-associated *pseudomonads* of worldwide origin. *FEMS Microbiology Ecology, 37*, 105–116.

Yang, S. F., & Hoffman, N. E. (1984). Ethylene biosynthesis and its regulation in higher plants. *Annual Review of Plant Physiology, 35*, 155–189.

Yuhashi, K. I., Ichikawa, N., Ezura, H., Akao, S., Minakawa, Y., Nukui, N., Yasuta, T., & Minamisawa, K. (2000). Rhizobitoxine production by *Bradyrhizobium elkanii* enhances nodulation and competitiveness on *Macroptilium atropurpureum*. *Applied and Environmental Microbiology, 66*, 2658–2663.

Eur J Plant Pathol (2007) 119:341–351
DOI 10.1007/s10658-007-9154-4

FULL RESEARCH PAPER

Effects of plant growth-promoting rhizobacteria on nodulation of *Phaseolus vulgaris* L. are dependent on plant P nutrition

Roseline Remans · Anja Croonenborghs · Roldan Torres Gutierrez · Jan Michiels · Jos Vanderleyden

Received: 8 December 2006 / Accepted: 26 April 2007 / Published online: 25 May 2007

Abstract Several plant growth-promoting rhizobacteria (PGPR) have shown potential to enhance nodulation of legumes when coinoculated with *Rhizobium*. To optimize the efficiency of these *Rhizobium*-PGPR-host plant interactions, unravelling the underlying mechanisms and analyzing the influence of specific environmental conditions is crucial. In this work the effect of four PGPR strains on the symbiotic interaction between *Rhizobium* and common bean (*Phaseolus vulgaris*) was studied under deficient versus sufficient phosphorus supply. It was observed that the effect on nodulation of three out of four PGPR tested was strongly dependent on P nutrition. Further, the use of specific PGPR mutant strains indicated that bacterial indole-3-acetic-acid production (IAA) and 1-aminocyclopropane-1-carboxylate (ACC) deaminase activity play an important role in the host nodulation response, particularly under low P conditions. Moreover, it was shown that the differential response to PGPR under low versus high P conditions was associated with changes in the host hormone sensitivity for nodulation induced under P deficiency. These findings contribute to the understanding of the interplay between *Rhizobium*, PGPR and the plant host under different environmental settings.

R. Remans (✉) · A. Croonenborghs ·
R. T. Gutierrez · J. Michiels · J. Vanderleyden
Department of Microbial and Molecular Systems, Centre of Microbial and Plant Genetics, K.U. Leuven, Kasteelpark Arenberg 20, 3001 Heverlee, Belgium
e-mail: roseline.remans@biw.kuleuven.be

Keywords Phosphorus deficiency · Nodulation · Common bean · Plant growth-promoting rhizobacteria · Phytohormones

Introduction

Legumes capable of symbiotic nitrogen fixation are crucial in agricultural crop production (Smith and Hume 1987; Vance 1997; Pepper 2000). Common bean (*Phaseolus vulgaris* L.) is the most important food legume for direct human consumption in the world (Broughton et al. 2003). In Africa and Latin America, common bean production is often undertaken on small farms using marginal soils low in nitrogen (N) and phosphorus (P) (Graham 1981; Thung 1991) and with minimal technical inputs. Progressive mining of soil nutrients does occur (Franzluebbers et al. 1998), with P second only to N as the most limiting factor for plant growth (Vance et al. 2000). P deficiency specifically is a nutritional limiting factor for efficient nodulation and nitrogen fixation (Christiansen and Graham 2002). N_2 fixation in legumes is a P-requiring process. Nodules of P-sufficient plants have a higher P concentration than shoots and roots, and may contain up to 1.5% of the total plant P (Graham and Rosas 1979; Pereira and

Bliss 1987, Schulze et al. 2006). Al-Niemi et al. (1997) suggest that nodule bacteroids in beans may be P limited even though the plant is well supplied with P. P-deficient plants exhibit reduced carbohydrate supply to nodules, and are usually restricted in nodule initiation, development and growth, and in nitrogenase activity (Graham and Rosas 1979; Pereira and Bliss 1987; Vadez et al. 1999; Kouas et al. 2005; Schulze et al. 2006).

Simultaneous inoculation with *Rhizobium* and other plant growth-promoting bacteria has shown potential to enhance plant growth, nodulation and nitrogen fixation of several legumes. Examples are numerous. Dual inoculation with *Azotobacter* spp. or *Azospirillum* spp. and *Rhizobium* strains showed a synergistic effect on nodulation, plant growth, yield and N uptake in soybean, clover, common bean and peanut (Burns et al. 1981; Raverker and Konde 1988; Burdman et al. 1997). Similarly, coinoculation with *Pseudomonas* spp. and *Rhizobium* spp. has been reported to enhance nodulation and nitrogen fixation, plant biomass and grain yield in various leguminous species including alfalfa (Bolton et al. 1990), soybean (Dashti et al. 1998), green gram (Sindhu et al. 1999) and chickpea (Goel et al. 2002). Coinoculation with *Bacillus* spp. and *Rhizobium* or *Bradyrhizobium* spp. ameliorated the nodulation and plant growth of common bean and soybean respectively (Srinivasan et al. 1997; Camacho et al. 2001; Bai et al. 2003). The mechanisms used by rhizobacteria to promote nodulation and nitrogen fixation are mostly unknown or remain ambiguous. Increased root development and nutrient uptake induced by plant growth-promoting rhizobacteria (PGPR) and their metabolites (reviewed by Lugtenberg et al. 2002; Persello-Cartieaux et al. 2003) may indirectly contribute to nodulation and nitrogen fixation (Vessey and Buss 2002). Evidence for a more direct effect on nodulation was described by Volpin et al. (1996) and Burdman et al. (1996). They observed that *A. brasilense* increased the exudation of flavonoids, plant signalling molecules crucial in the establishment of the *Rhizobium*-legume symbiosis, in the alfalfa and common bean rhizosphere. The use of a *gusA*-marked *Azospirillum* strain in an *Azospirillum–Rhizobium* coinoculation experiment on white clover demonstrated that *Azospirillum* colonized the tap root, root hairs and sites near or on the nodules (Tchebotar et al. 1998). Taken together with the observation that non-nodulated plants form more root hairs and lateral roots (Plazinski and Rolfe 1985a), it is suggested that when coinoculated with *Rhizobium*, *Azospirillum* stimulates the formation of epidermal cells that become infected root hair cells, or creates additional infection sites that are later occupied by rhizobia (Plazinski and Rolfe 1985c; Tchebotar et al. 1998). Recently, Thilak et al. (2006) showed that PGPR in conjunction with efficient *Rhizobium* can also affect the growth and nitrogen fixation in pigeon pea by enhancing the occupancy of introduced *Rhizobium* in the nodules of the legume. The nodule occupancy of the introduced *Rhizobium* strain increased from 50% (with *Rhizobium* alone) to 85% in the presence of *Pseudomonas putida*.

Positive effects of PGPR on nodulation, nitrogen fixation and plant growth of legumes have been observed in greenhouse experiments using hydroponic, vermiculite-based and soil-based systems as well as in field experiments (Burdman et al. 1997; Bai et al. 2003; Hamaoui et al. 2001). However, the influence of specific environmental factors on these *Rhizobium*-PGPR-plant interplays has not yet been studied. In the present work, the effect of different PGPR on the *Rhizobium*-bean symbiosis was evaluated under deficient versus sufficient P supply, thereby exploring the potential of *Rhizobium*-PGPR coinoculation to enhance nitrogen fixation under P deficiency. To initiate mechanistic studies of the synergistic effects, wild-type and specific mutants of some of the PGPR tested were used for the first time in *Rhizobium*-PGPR coinoculation experiments.

Materials and methods

Bacterial strains and growth conditions

The strains used in this study are listed in Table 1. *Rhizobium etli* CNPAF512 was grown overnight in liquid tryptone-yeast extract (TY) medium at 30°C or maintained on yeast extract -mannitol (YEM) (Vincent 1970) agar plates (15 g agar l^{-1}). *Azospirillum brasilense* Sp245 was grown overnight in liquid L* broth, which is Luria-Bertani broth medium (Sambrook et al. 1989) supplemented with 2.5 mM $CaCl_2$ and 2.5 mM $MgSO_4$, at 30°C or maintained on yeast extract-peptone agar plates (YEP)

Table 1 Bacterial strains used in this study

Bacterial strains	Relevant characteristics	Reference
Rhizobium etli CNPAF512	Wild-type strain, isolated from *Phaseolus vulgaris* nodules, Brazil	Michiels et al. (1998)
Azospirillum brasilense Sp245	Wild-type strain, isolated from surface-sterilized wheat roots, Brazil	Baldani et al. (1986)
FAJ009	*ipdC* Tn5 mutant of *A. brasilense* Sp245; Km^r	Costacurta et al. (1994)
Bacillus subtilis LMG7135	Wild-type strain, LMG bacteria collection	Chun and Bae (2000)
Pseudomonas putida UW4	Wild-type strain, isolated from the rhizosphere of reeds and selected for ACC deaminase activity, Canada	Glick et al. (1995)
$UW4/AcdS^-$	AcdS, ACC deaminase mutant of *P. putida*, Tc^r	Li et al. (2000)
Pseudomonas fluorescens SBW25	Wild-type SBW25	Rainey and Baily (1996)

(15 g agar l^{-1}). *Bacillus subtilis* LMG7135 was grown overnight in liquid TY medium at 30°C or maintained on TY agar plates (15 g agar l^{-1}). *Pseudomonas* strains, *Pseudomonas fluorescens* SBW25 and *Pseudomonas putida* UW4, were grown in liquid Tryptic Soy Broth (TSB) medium at 30°C or maintained on TSB agar plates (15 g agar l^{-1}). For the genetically modified strains the antibiotics kanamycin (Km) or tetracycline (Tc) were added to a final concentration of 25 and 10 μg ml^{-1} respectively.

Preparation of bacterial inocula

Pure cultures were grown overnight at 30°C in liquid medium (see above). Cells were washed twice with 10 mM $MgSO_4$ and resuspended in 10 mM $MgSO_4$ at a density of 10^7 colony forming units (cfu) ml^{-1} for *Rhizobium* or at a density of 10^9 cfu ml^{-1} for *Azospirillum*, *Bacillus* and *Pseudomonas* strains. *Rhizobium* inoculum contained only *Rhizobium* cells at a density of 10^7 cfu ml^{-1} $MgSO_4$. Coinoculants contained *Rhizobium* cells at a density of 10^7 cfu ml^{-1} and cells of other specified bacteria at a density of 10^9 cfu ml^{-1} $MgSO_4$.

Plant material and growth conditions

Seeds of *Phaseolus vulgaris* cvs BAT477 were kindly provided by the International Centre of Tropical Agriculture (CIAT), Cali, Colombia. A hydroponic system with a perlite-matrix was optimized to grow bean plants under low versus high P conditions in the greenhouse. One litre pots filled with perlite were initially moistened with 250 ml nitrogen-free medium (based on Snoeck et al. 2003). During growth a drip irrigation system maintained the pH (pH 6) and concentration of nutrients in the pots. The high P culture medium contained 100 μM KH_2PO_4, 1.71 mM K_2SO_4, 0.48 mM $MgSO_4$, 0.70 mM $MgCl_2$ and 1.06 mM $CaCl_2$. The low P medium was identical to the high P medium apart from the concentration of KH_2PO_4, which was 1 μM in low P conditions, and the concentration of K_2SO_4, which was 1.76 mM in low P medium (to maintain equal amounts of potassium under low versus high P conditions). An initial gift of 0.5 mmol KNO_3 was added to each pot at sowing.

The seeds of bean cv. BAT477 were surface-sterilized as described previously (Vlassak et al. 1998) and pre-germinated during two days on moist filter paper in the dark at 28°C. One pre-germinated seedling was planted per pot. The seedlings were inoculated with 200 μl inoculum (prepared as described above) containing 10^6 *Rhizobium* cells or 10^6 *Rhizobium* cells and 10^8 cells of *Azospirillum*, *Bacillus* or *Pseudomonas*. The number of cells used for inoculation is based on previous research described by Burdman et al. (1996), Burdman et al. (1997), Hamaoui et al. (2001) and Bai et al. (2002). For each condition 12 plants were used and arranged in a randomised block design. Plants were harvested at 21 and/or 42 days after inoculation (DAI) for determination of nodule number, nodule dry weight, shoot dry weight, root dry weight and NPK contents.

Hormone treatments

Low and high P culture media were supplemented with 0, 10, 100 or 200 nM indole-3-acetic acid (IAA). IAA was solubilised in ethanol at a concentration of 0.1 M before adding to the nutrient solution. IAA was purchased from Sigma Chemicals.

Plant analysis for NPK contents

Total N content (%N) was measured by ignition with a Variomax CN analyser. Total P and K contents (%P, %K) of shoot, root and nodule tissues were measured by Inductively Coupled Plasma—Optical Emission Spectroscopy (ICP-OES, Perkin Elmer, Optima 3300 DV, axial plasma).

Statistical analysis

Significance of difference between inoculation treatments and P levels was determined by the Tukey–Kramer Honestly Significant Difference Test at $P = 0.05$, using the SAS programme (SAS 9.1) (SAS Institute 1996). All experiments were performed at least twice, yielding similar results.

Results

Nodulation and plant growth parameters under deficient versus sufficient P conditions

BAT477 bean plants were inoculated with *Rhizobium etli* CNPAF512 and grown under low and high P conditions in the greenhouse. Plant growth and nodulation parameters were determined at harvest 42 DAI. The results, summarized in Table 2, show that (i) plants grown under low P (1 μM) conditions used in our experimental set-up are deficient in P; (ii) nodulation is more reduced than plant growth under P deficiency and (iii) nodules contain a higher concentration of P than other plant parts under high as well as under low P conditions. These results are consistent with previous observed effects of P deficiency on nodulation and growth parameters of common bean and other legumes.

Effect of PGPR on nodulation and plant growth parameters of bean under low versus high P conditions

BAT477 bean plants were inoculated with *Rhizobium etli* CNPAF512 or coinoculated with *Rhizobium* and other rhizobacteria, *Azospirillum brasilense* Sp245, *Bacillus subtilis* LMG7135, *Pseudomonas putida* UW4 or *Pseudomonas fluorescens* SBW25. For each of these rhizobacteria when used as a single inoculum plant growth-promoting activities have been extensively documented in the literature with various plant species. Similarly as described above, plants were grown under low and high P conditions in the greenhouse and harvested at 42 DAI. The effect of coinoculation on nodule number under high and low P conditions is shown in Fig. 1a, b respectively. In this figure the nodule number for each inoculation treatment is presented as % nodule number of plants inoculated with *Rhizobium* only. Under high P conditions, coinoculation with *Rhizobium* and *A. brasilense* Sp245 or with *Rhizobium* and *B. subtilis* LMG7135 significantly ($P < 0.05$) increased bean nodulation. However, under P deficient conditions, this positive effect of Sp245 and LMG7135 was not observed. On the contrary coinoculation with *A. brasilense* Sp245 resulted in a strong reduction of nodulation under low P. LMG7135 did not affect nodulation significantly under low P. The reverse effect of P nutrition was observed for coinoculation with *Rhizobium* and *P. putida* UW4. UW4 enhanced nodulation strongly under P deficiency but only slightly under P-sufficient conditions. *Pseudomonas fluorescens* SBW25 did not affect nodule number of bean line BAT477 significantly under the conditions tested.

Similar effects of the PGPR strains used were observed on nodule dry weight (data not shown), shoot dry weight (Fig. 1c, d) and root dry weight (Fig. 1e, f) under low versus high P conditions. The similarity in response to PGPR of nodule number, shoot and root dry weight strongly suggests a correlation between nodulation and plant growth parameters. Coinoculation with *Rhizobium* and PGPR did not affect plant NPK concentrations, only total NPK contents (data not shown).

Table 2 Nodulation and growth parameters of BAT477 bean plants inoculated with *Rhizobium etli* CNPAF512 and grown under low P (1 μM KH_2PO_4) and high P (100 μM KH_2PO_4) conditions

	High P	Low P	Low P As % of high P
Nodule number	359.2 ± 39	107.5 ± 18	29.8*
Nodule dry weight (mg)	200.4 ± 30.8	25.9 ± 5.6	12.9*
Acetylene reduction Activity (μl ethylene h^{-1})	7.1 ± 1.45	0 ± 0	0*
Shoot dry weight (SDW) (g)	2.06 ± 0.34	0.74 ± 0.18	35.9*
Root dry weight (RDW) (g)	0.51 ± 0.08	0.37 ± 0.05	72.6*
SDW/ RDW	4.04 ± 0.12	2.00 ± 0.09	49.5*
Shoot % N	3.86 ± 0.36	3.31 ± 0.25	85.8
Shoot % C	43.9 ± 0.44	42.4 ± 0.76	96.4
Shoot % P	0.22 ± 0.02	0.08 ± 0.02	36.4*
Shoot N/P	17.53 ± 1.06	41.40 ± 4.45	236.2*
Root % P	0.25 ± 0.03	0.16 ± 0.02	64.0*
Nodules % P	0.46 ± 0.05	0.34 ± 0.04	73.2*
Total N (mg)	146.5 ± 24.91	24.5 ± 6.56	16.7*
Total C (mg)	905.82 ± 121.7	313.64 ± 45.6	34.6*
Total P (mg)	4.53 ± 0.83	0.59 ± 0.12	13.0*

Values are averages of 12 plants ± standard deviations

*$P < 0.05$ Tukey–Kramer for difference between low P and high P values

Differential nodulation and growth response to PGPR under low versus high P is associated with production or conversion of phytohormones by PGPR

To delineate the mechanisms by which PGPR contribute to the differential nodulation response under low versus high P conditions, PGPR mutant strains affected in genes known to be involved in their plant growth-promoting effect were used. Bean plants were inoculated with *Rhizobium*, coinoculated with *Rhizobium* and a wild-type PGPR strain or with *Rhizobium* and a mutant PGPR strain, and grown during 42 days under low and high P conditions in the greenhouse. This allowed us to compare the effect of the mutant and wild-type PGPR strain on nodulation and to evaluate the role of the mutations in the modulation of nodulation.

Figure 2a, b shows that the effect of coinoculation on nodulation was less pronounced when the mutant strain FAJ009 of *A. brasilense* and the UW4/AcdS$^-$ mutant of *P. putida* UW4 were used instead of parent wild-type strains. FAJ009 is an insertion mutant of *A. brasilense* Sp245 in which the indole-3-pyruvate decarboxylase (*ipdC*) gene is inactivated, resulting in 90% reduction of production of the phytohormone indole-3-acetic acid in FAJ009 as compared to the wild-type Sp245 strain (Costacurta et al. 1994; Vande Broek et al. 1999). UW4/AcdS$^-$ (Li et al. 2000) is an insertion mutant of *P. putida* UW4 in which the gene encoding the enzyme 1-aminocyclopropane-1-carboxylate deaminase (ACCD) is inactivated. The enzyme 1-aminocyclopropane-1-carboxylate deaminase converts ACC, the precursor of the plant hormone ethylene, to alpha-ketobutyrate and ammonium and thereby reduces the production of ethylene. The mutations in FAJ009 and UW4/AcdS$^-$ have been described to be stable without antibiotic selection by Dobbelaere et al. (1999) for FAJ009 and by Hontzeas et al. (2004) for UW4/AcdS$^-$. In the media used no differences in growth rate were observed between wild-type and mutant strains.

Under low P conditions, FAJ009 did not affect nodulation significantly while the Sp245 wild-type strain reduced nodule number by 36% as compared to single *Rhizobium* inoculation. Under high P conditions, FAJ009 enhanced nodule number significantly, but the increase in nodulation was less pronounced compared to coinoculation with the wild-type strain.

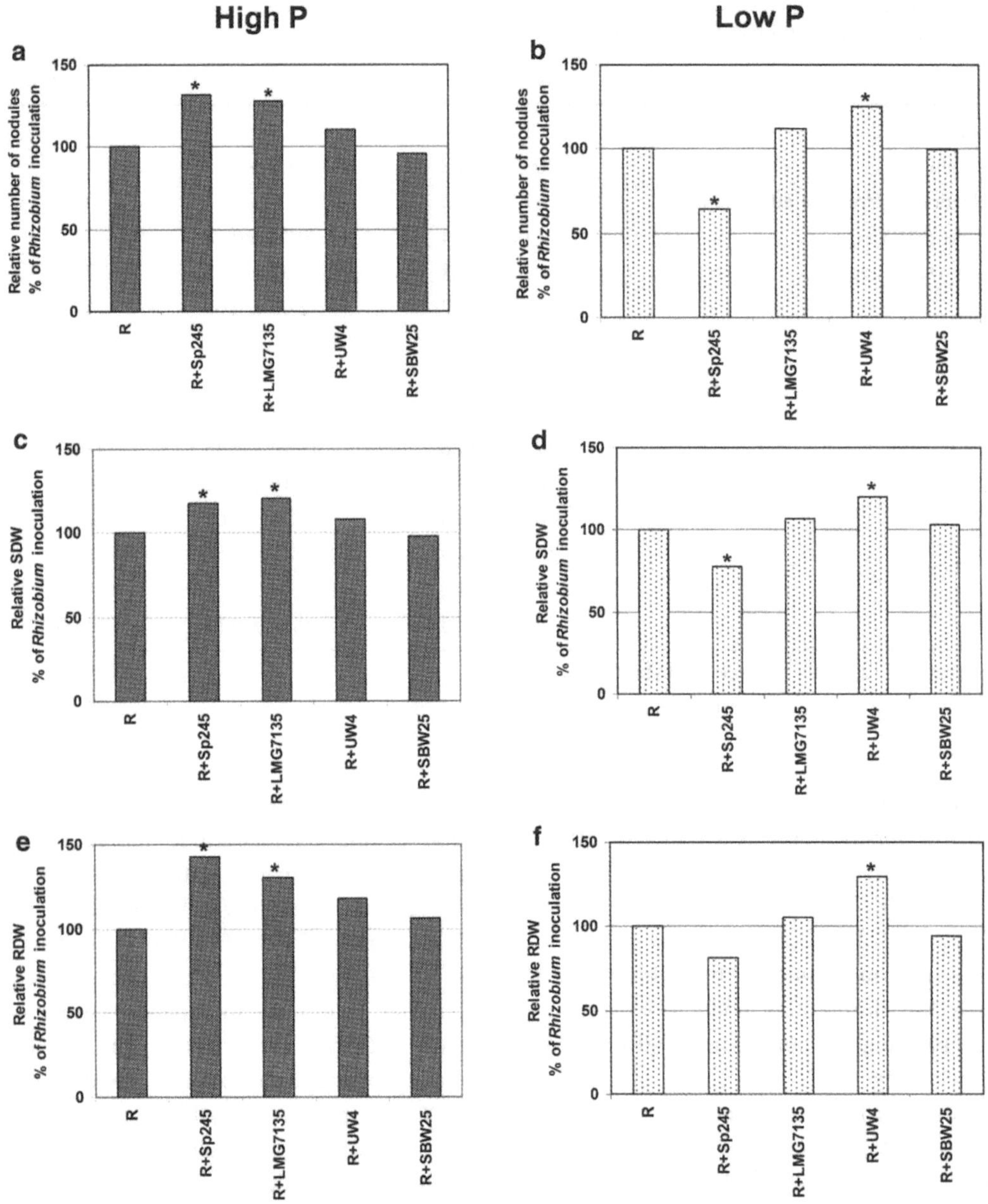

Fig. 1 Effect of PGPR on nodulation and plant parameters under high and low P conditions. Bars represent percentages of (**a**) nodule number per plant under high P; (**b**) nodule per plant number under low P; (**c**) plant shoot dry weight (SDW) under high P; (**d**) SDW under low P; (**e**) plant root dry weight (RDW) under high P (**f**) RDW under low P of BAT477 plants inoculated with *Rhizobium etli* CNPAF512 (R) or coinoculated with *Rhizobium* and *A. brasilense* Sp245 (R + Sp245), *B. subtilis* LMG7135 (R + LMG7135), *P. putida* UW4 (R + UW4), *P. fluorescens* SBW25 (R + SBW25). 100% represents the average value of plants inoculated with *Rhizobium* alone (see Table 1 for absolute values of nodule number, SDW, RDW). Plants were grown for 42 days under low and high P conditions and harvested for nodulation and plant parameters analysis. 12 plants per condition were used. $*P < 0.05$ Tukey–Kramer. This experiment was performed three times, yielding similar results

Inactivation of the ACC-deaminase in *P. putida* UW4 reduced the positive effect on nodulation of UW4 under low P conditions. Under high P conditions, no significant differences were observed between single *Rhizobium* inoculation and coinoculation with *Rhizobium* and UW4 or its ACCD minus mutant.

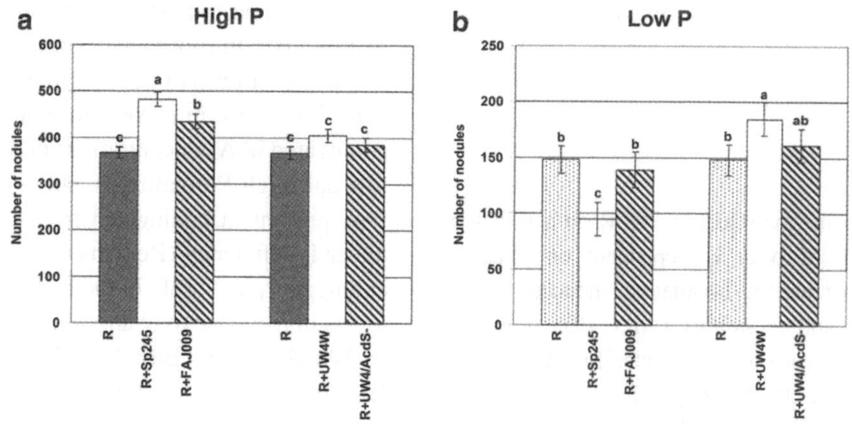

Fig. 2 The effect on nodule number of mutant strains of PGPR compared to the effect of their wild-type strain under (**a**) high P and (**b**) low P conditions. Bars represent average values of nodule number of BAT477 bean plants inoculated with *Rhizobium etli* CNPAF512 (R), coinoculated with *Rhizobium* and the wild-type strain of *A. brasilense* Sp245 (R + Sp245 WT, white bars) or the wild-type strain of *P. putida* UW4 (R + UW4 WT, white bars), or coinoculated with *Rhizobium* and the FAJ009 mutant of Sp245 (R + FAJ009, striped bars) or the AcdS$^-$ mutant of UW4 (R + UW4/AcdS$^-$, stride bars). Plants were grown for 42 days under low and high P conditions and harvested for nodulation and plant parameters analysis. 12 plants per condition were used. Error bars represent ±95% confidence intervals (Tukey–Kramer). Letters on top of the bars indicate different classes of statistical differences (Tukey–Kramer). This experiment was performed three times, yielding similar results

Effect of IAA on nodulation under low versus high P conditions

To further unravel the role of exogenous IAA in modulation of nodulation under low and high P conditions, increasing concentrations of IAA were added to the nutrient solution medium of bean plants inoculated with *Rhizobium*. BAT477 plants were grown for 21 days under low and high P conditions and harvested for analysis of early nodulation and plant growth. Figure 3 shows the dose-response curve of nodule number per plant (represented as % of average nodule number at 0 nM IAA) to increasing concentrations of IAA in the nutrient solution. It was observed that IAA at low concentrations enhanced nodule number, while higher concentrations inhibited nodulation under high as well as under low P conditions. These results underline the dose-dependent effect of IAA on nodulation. This is consistent with what has been described previously for other legumes (Van Noorden et al. 2006; Plazinski and Rolfe 1985b, c). Interestingly, nodulation was more responsive to IAA under P deficiency than under P-sufficient conditions. Under low P conditions, a concentration of 100 nM IAA already strongly inhibited nodulation, while the same concentration

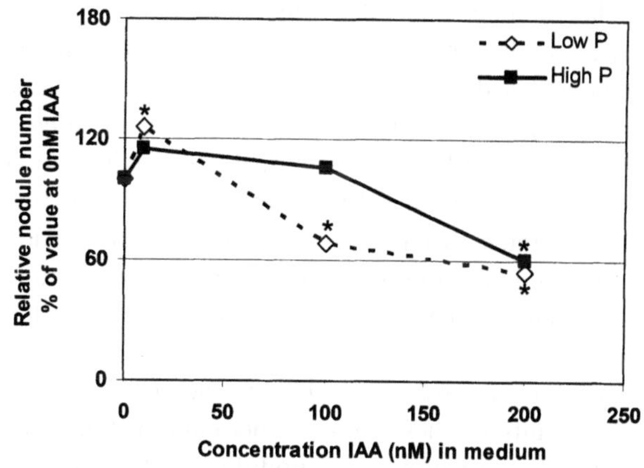

Fig. 3 The effect on nodule number of increasing concentrations of indole-3-acetic acid (IAA) under high P versus low P conditions. Points represent percentages of nodule number of BAT477 bean plants inoculated with *Rhizobium etli* CNPAF512 and grown in the presence of different concentrations of indole-3-acetic acid (IAA) under low P (dashed line) and high P (full line) conditions. 100% represents the average nodule number per plant grown at 0 nM IAA. Under high P 100% is equal to 142 nodules per plant; under low P 100% is equal to 85 nodules per plant. Plants were harvested 21 days after inoculation. IAA was added to the nutrient solution as described in materials and methods. * indicates significant difference ($P < 0.05$ Tukey–Kramer) for absolute values between plants grown in the presence of IAA and control plants grown in the absence of IAA (0 nM). This experiment was performed three times, yielding similar results

of IAA under high P conditions still slightly enhanced nodule number.

Discussion

This study shows that the effect of PGPR on nodulation and plant growth of bean is dependent on plant P nutrition. The increase in nodulation induced by *A. brasilense* Sp245 and *B. subtilis* LMG 7135 observed under high P, was not observed under low P. On the contrary, the positive effect of *P. putida* UW4 was more strongly expressed under P deficiency than under sufficient P supply. These results encourage integration of environmental factors including P nutrition in studies exploring the potential of *Rhizobium*-PGPR coinoculation to enhance nodulation and plant growth.

The difference in plant and nodulation response under low versus high P conditions was most pronounced when plants were inoculated with the combination of *Rhizobium* and *A. brasilense* Sp245. Under high P conditions this combination enhanced nodule number, shoot dry weight and root dry weight significantly, while under low P the same inoculation treatment had a negative effect on these parameters. Intriguingly, using an *ipdC* mutant of *A. brasilense* Sp245 reduced IAA biosynthesis by 90% (Costacurta et al. 1994; Vande Broek et al. 1999); it was demonstrated that the *Azospirillum* IAA production plays a crucial role in the modulation of nodulation, particularly under low P conditions. This was further supported by experiments in which increasing concentrations of IAA were added to the growth media. The results of these experiments confirm the dose-dependent response of nodulation to IAA as described previously (Van Noorden et al. 2006; Plazinski and Rolfe 1985a, b). Low concentrations (up to 100 nM IAA) can enhance nodule number, while higher concentrations inhibit nodulation. Interestingly, Plazinski and Rolfe (1985a, b) described a similar dose-response curve of nodulation in response to inoculation with an increasing number of *Azospirillum* cells on bean plants.

The experiments using increasing IAA concentrations further showed that nodulation was more sensitive to IAA under low P than under high P conditions. Increase in auxin sensitivity under P-deficient conditions has previously been described in *Arabidopsis* (Lopez-Bucio et al. 2002). The use of *Arabidopsis* auxin-resistant mutants also demonstrated the crucial role of auxin in root morphological adaptations under low P conditions (Lopez-Bucio et al. 2002; Al-Ghazi et al. 2003). Preliminary data (not shown) of free and conjugated IAA concentrations in BAT477 roots under low and high P conditions, show that the amount of IAA present in conjugated form is strongly reduced under P deficiency (Remans et al. unpublished). IAA conjugates play a role in the maintenance of free IAA levels in response to changes in IAA levels (reviewed by Woodward and Bartel 2005), e.g. when exogenous IAA is added, the difference in root IAA conjugate concentrations between low and high P conditions further supports the observation of increased IAA sensitivity under low P deficiency. Since IAA sensitivity and adaptation to P deficiency is strongly dependent on the plant genotype, the response to *A. brasilense* Sp245 may also differ strongly across contrasting bean genotypes.

Apart from changes in host sensitivity to IAA, changes in IAA production rate induced in *A. brasilense* Sp245 under low P conditions, may also contribute to the differential response in nodulation to Sp245 inoculation under low versus high P conditions. It has been described that IAA biosynthesis in *A. brasilense* increases under certain environmental stress conditions, including low pH and carbon stress (Ona et al. 2005). The specific effect of P deficiency on IAA production in Sp245 has not yet been studied. The negative effect of IAA on plant growth and nodulation is often mediated by ethylene (Persello-Cartieaux et al. 2003). Auxin produced by bacteria in the rhizosphere can stimulate the activity of the 1-aminocyclopropane-1-carboxylate (ACC) synthase, an enzyme used by plants for the synthesis of ethylene (Xie et al. 1996). Ethylene has been shown to negatively regulate the plant's response to the rhizobial bacterial signal, Nod factor (Oldroyd et al. 2001; Penmetsa and Cook 1997). The inhibition of Nod factor-induced calcium spiking reflects the suppression of the Nod factor signalling pathway at a very early stage and most probably explains the regulation of nodule number by ethylene. Borch et al. (1999) reported that P-deficient roots of common bean produced twice as much ethylene g^{-1} root dry weight than P-sufficient roots. The increased ethylene production in P-deficient roots may contribute to the reduction in nodulation under low P conditions. Interestingly, the positive effect of coinoculation with *Pseudomonas putida* UW4 on nodulation was most expressed under low P conditions.

This enhancement in nodulation was much less pronounced when an UW4 mutant disabled in ACC deaminase activity was used. These results indicate that bacterial ACC deaminase activity, which reduces ethylene production, may play a crucial role in increasing nodulation and plant growth under P deficiency. Direct evidence for a positive effect of ACC deaminase activity on nodulation of alfalfa has been described previously by Ma et al. (2004). They showed that the expression of an exogenous ACC deaminase gene in *Sinorhizobium meliloti* increases its ability to nodulate alfalfa.

In conclusion, this study has showed that P deficiency alters the effect of PGPR on nodulation and growth of common bean. The differential nodulation response to PGPR under low versus high P conditions is associated with changes in phytohormone balances induced by specific PGPR. Potential to enhance nodulation and plant growth under P deficiency by coinoculation with *Rhizobium* and PGPR does exist, particularly for bacteria possessing ACC deaminase activity. Further research to fine-tune combinations of *Rhizobium*, PGPR and host genotype will ultimately lead to better nodulation and stimulated plant growth under low P conditions.

Acknowledgements R.R. is a recipient of a predoctoral fellowship from the 'Vlaamse Interuniversitaire Raad (VLIR)' R.T. acknowledges the receipt of a predoctoral fellowship from the Katholieke Universiteit Leuven. We thank Dr. Bernard Glick for kindly providing us the strain *P. putida* UW4 and strain UW4/AcdS^{-}.

References

Al-Ghazi, Y., Muller, B., Pinloche, S., Tranbarger, T. J., Rossignol, M., Tardieu, F., & Doumas, P. (2003). Temporal responses of *Arabidopsis* root architecture to phosphate starvation: Evidence for the involvement of auxin signaling. *Plant, Cell and Environment, 26*, 1053–1066.

Al-Niemi, T. S., Kahn, M. L., & McDermott, T. R. (1997). P metabolism in the bean-*Rhizobium*-tropici symbiosis. *Plant Physiology, 113*, 1233–1242.

Bai, Y., Panb, B., Charlesc, T., & Smith, D. (2002). Co-inoculation dose and root zone temperature for plant growth promoting rhizobacteria on soybean [*Glycine max* (L.) Merr] grown in soil-less media. *Soil Biology and Biochemistry, 34*, 1953–1957.

Bai, Y., Zhou, X., & Smith, D. (2003). Enhanced soybean plant growth resulting from coinoculation of *Bacillus* strains with *Bradyrhizobium japonicum*. *Crop Science, 43*, 1174–1781.

Baldani, V. L. D., Alvarez, M. A. B., Baldani, J. I., & Döbereiner, J. (1986). Establishment of inoculated *Azospirillum* spp. in the rhizosphere and in roots of field grown wheat and sorghum. *Plant Soil, 90*, 37–40.

Bolton, H., Elliott, L. F., Turco, R. F., & Kennedy, A. C. (1990). Rhizoplane colonization of pea seedlings by *Rhizobium leguminosarum* and a deleterious root colonizing *Pseudomonas* sp. and effects on plant growth. *Plant Soil, 123*, 121–124.

Borch, K., Bouma, T. J., Lynch, J. P., & Brown, K. M. (1999). Ethylene: A regulator of root architectural responses to soil phosphorus availability. *Plant Cell Environment, 22*, 425–431.

Broughton, W. J., Hernander, G., Blair, B., Beebe, S., Gepts, P., & Vanderleyden, J. (2003). Beans (*Phaseolus* spp.) – model food legumes. *Plant Soil, 252*, 55–128.

Burdman, S., Kigel, J., & Okon, Y. (1997). Effects of *Azospirillum brasilense* on nodulation and growth of common bean (*Phaseolus vulgaris* L.). *Soil Biology and Biochemistry, 29*, 923–929.

Burdman, S., Volpin, H., Kigel, J., Kapulnik, Y., & Okon, Y. (1996). Promotion of nod gene inducers and nodulation in common bean (*Phaseolus vulgaris*) roots inoculated with *Azospirillum brasilense* Cd. *Applied and Environmental Microbiology, 62*, 3030–3033.

Burns, T. A., Bishop, P. E., & Israel, D. W. (1981). Enhanced nodulation of leguminous plant roots by mixed cultures of *Azotobacter vinelandi* and damping-off of tomato by *Pseudomonas aeruginosa* 7NSK2. *Applied and Environmental Microbiology, 62*, 865–871.

Camacho, M., Santamaria, C., Temprano, F., Rodriguez-Navarro, D. N., & Daza, A. (2001). Co-inoculation with Bacillus sp. CECT 450 improves nodulation in *Phaseolus vulgaris* L. *Canadian Journal of Microbiology, 47*, 1058–1062.

Christiansen, I., & Graham, P. H. (2002). Variation in di-nitrogen fixation among Andean bean (*Phaseolus vulgaris* L.) genotypes grown at low and high levels of phosphorus supply. *Fields Crops Research, 73*, 133–142.

Chun, J., & Bae, K. S. (2000). Phylogenetic analysis of *Bacillus subtilis* and related taxa based on partial gyrA gene sequences. *Antonie Leeuwenhoek, 78*, 123–127.

Costacurta, A., Keijers, V., & Vanderleyden, J. (1994). Molecular cloning and sequence analysis of an *Azospirillum brasilense* indole-3-pyruvate decarboxylase gene. *Molecular and General Genetics, 243*, 463–472.

Dashti, N., Zhang, F., Hynes, R., & Smith, D. L. (1998). Plant growth promoting rhizobacteria accelerate nodulation and increase nitrogen fixation activity by field grown soybean [*Glycine max* (L.) Merr.] under short season conditions. *Plant Soil, 200*, 205–213.

Dobbelaere, S., Croonenborghs, A., Thys, A., Vande Broek, A., & Vanderleyden, J. (1999). Phytostimulatory effect of *Azospirillum brasilense* wild type and mutant strains altered in IAA production on wheat. *Plant Soil, 212*, 155–164.

Franzluebbers, K., Hossner, L. R., & Juo, A. S. R. (1998). Integrated nutrient management for sustained crop production in sub-Saharan agriculture. Trop Soils/TAMU Technical Bulletin No. 98-03, 50 pp.

Glick, B. R., Karaturovíc, D. M., & Newell, P. C. (1995). A novel procedure for rapid isolation of plant growth-promoting pseudomonads. *Canadian Journal of Microbiology, 41*, 533–536.

Goel, A. K., Sindhu, S. S., & Dadarwal, K. R. (2002). Stimulation of nodulation and plant growth of chickpea (*Cicer arietnum* L.) by *Pseudomonas* spp. antagonistic to fungal pathogens. *Biology and Fertility of Soils, 36*, 391–396.

Graham, P. H. (1981). Some problems of nodulation and symbiotic nitrogen fixation in *Phaseolus vulgaris* L.: A review. *Field Crops Research, 4*, 93–112.

Graham, P. H., & Rosas, J. C. (1979). Phosphorus fertilization and symbiotic nitrogen fixation in common bean. *Agronomy Journal, 71*, 925–926.

Hamaoui, B., Abbadi, J. M., Burdman, S., Rashid, A., Sarig, S., & Okon, Y. (2001). Effects on inoculation with *Azospirillum brasilense* on chickpeas (*Cicer arietum*) and faba beans (*Vicia faba*) under different growth conditions. *Agronomie, 21*, 553–560.

Hontzeas, N., Saleh, S. S., & Glick, B. R. (2004). Changes in gene expression in canola roots induced by ACC-deaminase-containing plant-growth-promoting bacteria. *Molecular Plant Microbe Interactions, 17*, 865–871.

Kouas, S., Labidi, N., Debez, A., & Abdelly, C. (2005). Effect of P on nodule formation and N fixation in bean. *Agronomy for Sustainable Development, 25*, 389–393.

Li, J., Ovakim, D. H., Charles, T. C., & Glick, B. R. (2000). An ACC deaminase minus mutant of *Enterobacter cloacae* UW4 no longer promotes root elongation. *Current Microbiology, 41*, 101–105.

Lopez-Bucio, J., Hernandez-Abreu, E., Sanchez-Calderon, L., Nieto-Jacobo, M. F., Simpson, J., & Herrera-Estrella, L. (2002). Phosphate availability alters architecture and causes changes in hormone sensitivity in the *Arabidopsis* root system. *Plant Physiology, 129*, 244–256.

Lugtenberg, B. J. J., Chin-A-Woeng, T. F. C., & Bloemberg, G. V. (2002). Microbe-plant interactions: Principles and mechanisms. *Antonie Leeuwenhoek, 81*, 373–383.

Ma, W., Charles, T. C., & Glick, B. R. (2004). Expression of an exogenous 1-aminocyclopropane-1-carboxylate deaminase gene in *Sinorhizobium meliloti* increases its ability to nodulate alfalfa. *Environmental Microbiology, 70*, 5891–5897.

Michiels, J., Moris, M., Dombrecht, B., Verreth, C., & Vanderleyden, J. (1998). Differential regulation of *Rhizobium etli* rpoN2 gene expression during symbiosis and free-living growth. *Journal of Bacteriology, 180*, 3620–3628.

Oldroyd, G. E., Engstrom, E. M., & Long, S. R. (2001). Ethylene inhibits the Nod factor signal transduction pathway of *Medicago truncatula*. *Plant Cell, 13*, 1835–1849.

Ona, O., Van Impe, J., Prinsen, E., & Vanderleyden, J. (2005). Growth and indole-3-acetic acid biosynthesis of *Azospirillum brasilense* Sp245 is environmentally controlled. *FEMS Microbiology Letters, 246*, 125–132.

Penmetsa, R. V., & Cook, D. R. (1997). A legume ethylene-insensitive mutant hyperinfected by its rhizobial symbiont. *Science, 257*, 527–530.

Pepper, J. L. (2000). Beneficial and pathogenic microbes in agriculture. In R. M. Maier et al. (Eds.), *Environmental microbiology*. San Diego: Academic Press.

Pereira, P. A., & Bliss, F. A. (1987). Nitrogen fixation and plant growth of common bean (*Phaseolus vulgaris* L.) at different levels of phosphorus availability. *Plant Soil, 104*, 79–84.

Persello-Cartieaux, F., Nussaume, L., & Robaglia, C. (2003). Tales from the underground: Molecular plant-rhizobacteria interactions. *Plant, Cell and Environment, 26*, 189–199.

Plazinski, J., & Rolfe B. G. (1985a). *Azospirillum–Rhizobium* interaction leading to a plant growth stimulation without nodule formation. *Canadian Journal of Microbiology, 31*, 1026–1030.

Plazinski, J., & Rolfe, B. G. (1985b). Interaction of *Azospirillum* and *Rhizobium* strains leading to inhibition of nodulation. *Applied and Environmental Microbiology, 49*, 990–993.

Plazinski, J., & Rolfe, B. G. (1985c). Influence of *Azospirillum* strains on the nodulation of clovers by *Rhizobium* strains. *Applied and Environmental Microbiology, 49*, 984–989.

Rainey, P. B., & Bailey, M. J. (1996). Physical map of the *Pseudomonas fluorescens* SBW25 chromosome. *Molecular Microbiology, 19*, 521–533.

Raverker, K. P., & Konde, B. K. (1988). Effect of *Rhizobium* and *Azospirillum lipoferum* inoculation on nodulation, yield and nitrogen uptake of peanut cultivars. *Plant Soil, 106*, 249–252.

Sambrook, J., Fritsch, E. F., & Maniatis, T. (1989). *Molecular cloning: A laboratory manual* (2nd ed.). Cold Spring Harbor, N.Y.: Cold Spring Harbor Laboratory Press.

SAS Institute (1996). *SAS user's guide: Statistics* (6th ed.). Cary, NC: SAS Institute.

Schulze, J., Temple, G., Temple, S. J., Beschow, H., & Vance, C. P. (2006). Nitrogen fixation by white lupin under phosphorus deficiency. *Annals of Botany, 98*, 731–740.

Sindhu, S. S., Gupta, S. K., & Dadarwal, K. R. (1999). Antagonistic effect of *Pseudomonas* spp. on pathogenic fungi and enhancement of plant growth in green gram (*Vigna radiata*). *Biology and Fertility of Soils, 29*, 62–68.

Smith, D. L., & Hume, D. J. (1987). Comparison of assay methods for nitrogen fixation utilizing white bean and soybean. *Canadian Journal of Plant Science, 67*, 11–79.

Snoeck, C., Verreth, C., Hernandez-Lucas, I., Martínez-Romero, E., & Vanderleyden, J. (2003). Identification of a third sulfate activation system in *Sinorhizobium* sp. strain BR816: The CysDN sulfate activation complex. *Applied and Environmental Microbiology, 69*, 2006–2014.

Srinivasan, M., Petersen, D. J., & Holl, F. B. (1997). Influence of indoleacetic-acid-producing *Bacillus* isolates on the nodulation of *Phaseolus vulgaris* by *Rhizobium etli* under gnobiotic conditions. *Canadian Journal of Microbiology, 42*, 1006–1014.

Tchebotar, V. K., Kang, U. G., Asis, C. A., & Akao, J. S. (1998). The use of GUS-reporter gene to study the effect of *Azospirillum–Rhizobium* coinoculation on nodulation of white clover. *Biology and Fertility of Soils, 27*, 349–352.

Thilak, K. V. B. R., Ranganayaki, N., & Manoharachari, C. (2006). Synergistic effects of plant-growth promoting rhizobacteria and *Rhizobium* on nodulation and nitrogen fixation by pigeonpea (*Cajanus cajan*). *European Journal of Soil Science, 57*, 67–71.

Thung, M. (1991). Bean agronomy in monoculture. In A. V. Schoonhoven & O. Voysest (Eds.), *Common beans. Research for crop improvement* (pp. 737–834). Wallingford, UK: CAB International.

Vadez, V., Lasso, J. H., Beck, D. P., & Drevon, J. J. (1999). Variability of N_2-fixation in common bean (*Phaseolus vulgaris* L.) under P deficiency is related to P use efficiency. *Euphytica, 106*, 231–242.

Vance, C. P. (1997). Enhanced agricultural sustainability through biological nitrogen fixation. In A. Legocki et al. (Eds.), *Biological fixation of nitrogen for ecology and sustainable agriculture*. Heidelberg, Germany: NATO ASI Series, Springer-Verslag.

Vance, C. P., Graham, P. H., & Allan, D. L. (2000). Biological nitrogen fixation: Phosphorus-a critical need. In F. A. Pedrosa et al. (Eds.), *Nitrogen fixation: From molecules to crop productivity* (pp. 509–514). Netherlands: Kluwer Academic Publishers.

Vande Broek, A., Lambrecht, M., Eggermont, K., & Vanderleyden, J. (1999). Auxins upregulate expression of the indole-3-pyruvate decarboxylase gene in *Azospirillum* brasilense. *Journal of Bacteriology, 181*, 1338–1342.

Van Noorden, G. E., Ross, J. J., Reid, J. B., Rolfe, B. G., & Mathesius, U. (2006). Defective long distance auxin transport regulation in the *Medicago truncatula* super numeric nodules mutant. *Plant Physiology, 140*, 1494–1506.

Vessey, K., & Buss, T. J. (2002). Bacillus cereus UW85 inoculation effects on growth, nodulation, and N accumulation in grain legumes – Controlled-environment studies. *Canadian Journal of Plant Science, 82*, 282–290.

Vincent, J. M. (1970). *A manual for the practical study of root-nodule bacteria*. Oxford: Blackwell Scientific Publishers.

Vlassak, K. M., Luyten, E., Verreth, C., van Rhijn, P., Bisseling, T., & Vanderleyden, J. (1998). The *Rhizobium* sp. BR816 *nodO* gene can function as a determinant for nodulation of *Leucaena leucocephala*, *Phaseolus vulgaris* and *Trifolium repens* by a diversity of *Rhizobium* spp. *Molecular Plant Microbe Interactions, 5*, 383–392.

Volpin, H., Burdman, S., Castro-Sowinski, S., Kapulnik, Y., & Okon, Y. (1996). Inoculation with *Azospirillum* increased exudation of Rhizobial nod-gene inducers by alfalfa roots. *Molecular Plant Microbe Interactions, 1996*, 388–394.

Woodward, A. W., & Bartel, B. (2005). Auxin: Regulation, action and interaction. *Annals of Botany, 95*, 707–735.

Xie, H., Pasternak, J. J., & Glick, B. R. (1996). Isolation and characterization of mutants of the plant growth-promoting rhizobacterium *Pseudomonas putida* CR12-2 that overproduce indoleacetic acid. *Current Microbiology, 32*, 67–71.

Eur J Plant Pathol (2007) 119:353–365
DOI 10.1007/s10658-007-9149-1

REVIEW PAPER

Quorum sensing as a target for developing control strategies for the plant pathogen *Pectobacterium*

Denis Faure · Yves Dessaux

Received: 19 December 2006 / Accepted: 19 April 2007 / Published online: 19 June 2007

Abstract Quorum sensing is a regulatory mechanism that connects gene expression to cell density in bacteria. Amongst proteobacteria, numerous functions are regulated in this way, including pathogenicity in the *Enterobacteriaceae* genus *Pectobacterium*. In *Pectobacterium*, the signalling molecules involved in this regulatory process belong to the N-acyl-homoserine lactone class. Over the last 6 years, various studies have shown that these signal molecules could be degraded by other bacteria or by plant and animal cells, opening the path to innovative biocontrol strategies. This review explores the various determinants of pathogenicity in *Pectobacterium* and describes approaches that have been developed to quench the quorum-sensing-dependent pathogenicity in *Pectobacterium*. These approaches range from signal degradation by physicochemical constraints to the identification of signal-sensing inhibitors and from the identification of enzymes degrading acyl-homoserine lactones to the construction of transgenic plants tolerant to *Pectobacterium*.

Keywords *Erwinia* · *Pectobacterium* · N-acyl-homoserine lactone · Quorum · Quenching · Virulence

D. Faure · Y. Dessaux (✉)
Institut des Sciences du Végétal, CNRS, Avenue de la terrasse, 91198 Gif-sur-Yvette Cedex, France
e-mail: dessaux@isv.cnr-gif.fr

Abbreviations

3O,C6-HSL	3-Oxo,hexanoyl-homoserine lactone
3O,C12-HSL	3-Oxo,dodecanoyl-homoserine lactone
C4-HSL	Butyroyl-homoserine lactone
C6-HSL	Hexanoyl-homoserine lactone
C12-HSL	Dodecanoyl-homoserine lactone
C14-HSL	Tetradecanoyl-homoserine lactone
GABA	γ-Aminobutyrate
GBL	γ-Butyrolactone
GHB	γ-Hydroxybutyrate
SSA	Succinic semialdehyde
N-AHSL	N-acyl-homoserine lactone
PCWME	Plant-cell-wall-macerating enzymes
PON	Paraoxonase
QS	Quorum sensing
QQ	quorum quenching
SAM	S-adenosyl-methionine
TTSS	Type III secretion system

Introduction

Quorum sensing (QS) regulation

Bacteria have evolved sophisticated mechanisms to coordinate gene expression at population and community levels. For instance, gene expression may depend upon the perception of diffusible molecules that are synthesized by bacterial populations and communities. Because the concentration of the emitted signal in a confined environment reflects the bacterial cell number and density, such a regulatory pathway was termed 'quorum sensing' (QS) (Fuqua et al. 1994). In an open environment, however, the concentration of the signal reflects the bacterial cell number and the signal diffusion coefficient. In such open environments, the term 'diffusion sensing' was proposed (Redfield 2002). Specific signal sensors and transcriptional factors, the activities of which are modulated by the concentrations of the emitted signals, are involved in QS.

The signal molecules

The structures of QS signals are highly diverse (Whitehead et al. 2001; Waters and Bassler 2005). Oligopeptides and substituted gamma-butyrolactones have been described in Gram-positive bacteria, while other substituted gamma-butyrolactones, the N-acyl-homoserine lactones (N-AHSLs), are synthesized by a large number of Gram-negative bacteria. In this latter bacterial group, 3-hydroxypalmitic acid methyl ester (Flavier et al. 1997), 3,4-dihydroxy-2-heptyl-quinoline (Holden et al. 1999), and a furanosyl borate diester (Chen et al. 2002) can also act as QS signals. Among Gram-negative bacteria, the most common QS signals are N-AHSL (Greenberg 2000; Fuqua et al. 2001; Whitehead et al. 2001). The synthesis of N-AHSL depends upon synthases belonging generally to two classes: the LuxI homologs and the AinS homologs (Fuqua and Greenberg 2002). The perception of the signal relies upon a sensor protein, a LuxR homolog, which is also the transcriptional regulator controlling the expression of QS-regulated genes (Fuqua and Greenberg 1998).

Quorum quenching (QQ)

The term quorum quenching (QQ) encompasses various natural phenomena or engineered procedures that lead to the perturbation and—eventually—the attenuation of the expression of QS-regulated functions (for recent reviews see: Dong and Zhang 2005; Rasmussen and Givskov 2006). Three main steps of the QS regulation could be targeted: the signal synthesis, the stability of signal, or the sensing of the signal. So far, QQ strategies have only dealt with the production, the accumulation or the perception of signals belonging to the N-AHSL class. In theory, however, similar strategies could be developed for QS processes relying upon other molecules. Because QQ targets the expression of virulence functions and does not affect the viability of bacterial pathogens, QQ falls into the family of anti-virulence/anti-disease strategies.

The synthases as targets

Few compounds have been identified as potential inhibitors of bacterial N-AHSL production. N-AHSL synthesis proceeds from S-adenosyl-methionine (SAM) and a fatty acid, linked to an acyl carrier protein. Amongst the synthase inhibitors, the two SAM analogues L-*S*-adenosylhomocysteine and sinefugin (an S-adenosyl-methionine-like antibiotic; Geze et al. 1983) were the most efficient (Parsek et al. 1999). However, other potential targets exist, such as proteins implicated in the synthesis of N-AHSL precursors, SAM or fatty acids. In agreement with this suggestion, mutants of *P. syringae* pv. *tabacci* affected in the acyl-(acyl carrier protein) synthase exhibited a phenotype similar to a *luxI* mutant (Taguchi et al. 2006). The bactericidal molecule triclosan—targeting the enoyl-acyl carrier protein reductase FabI—also affects the synthesis of N-AHSL (Hoang and Schweizer 1999). Whatever the target, this type of approach affects key metabolic compounds in bacteria. It is therefore likely to impair both QS-regulated functions and functions other than those regulated by QS in bacteria, a major drawback in the development of potential, specific inhibitors. Such bactericidal compounds therefore cannot be categorized as anti-virulence molecules.

The sensor as a target

The regulatory proteins of the LuxR family that senses N-AHSL have also been proposed as potential targets for QQ. Such a mechanism occurs in nature. For instance, the red alga *Delisea pulchra* limits bacterial colonization (fouling) by interfering with the QS-controlled motility and biofilm-formation ability of bacteria (Rasmussen et al. 2000). This process is mediated by halogenated furanones produced by the algae (Givskov et al. 1996). These molecules bind the LuxR receptor of potential bacterial colonizers, prevent the binding or displace the N-AHSL signal (Manefield et al. 1999), and accelerate the degradation of the LuxR protein (Manefield et al. 2002). Similar phenomena have been observed in another alga, *Chlamydomonas reinhardtii*, that produces over a dozen compounds which, most likely, are not furanones and inactivate N-AHSL-mediated QS functions in bacteria (Teplitski et al. 2004). Other inhibitors have been found in plants and more generally in bioproducts. Thus, pea and soybean (Teplitski et al. 2000), *Medicago* (Gao et al. 2003), fruit extracts such as those from grape and strawberry (Fray 2002), garlic (Rasmussen et al. 2005a), vanilla (Choo et al. 2006), lily and pepper (Rasmussen and Givskov 2006), *Clematis vitalba*, *Geranium molle*, and *Tropaeolum majusi* (Karamanoli and Lindow 2006) produce molecules that inhibit QS in bacteria. So far, only a few active molecules have been identified. In garlic, disulfur compounds with QS-antagonistic activity have been reported (Rasmussen and Givskov 2006). In other plant extracts, as in *D. pulchra*, furanones may be involved in the inhibition of QS. These molecules are major constituents of the fruity or spicy aromas of several plant products (Colin Slaughter 1999).

Fungi also produce inhibitors of QS. A screen of 50 *Penicillum* species revealed that about 50% produced inhibitors, two of these being identified as the lactones patulin and penicillic acid (Rasmussen et al. 2005b). Interestingly, patulin naturally occurs in fruits such as apple, pear, peach, apricot, banana, pineapple, and grape (Scott et al. 1972; Frank 1977), where the compound may also contribute to the inhibition of QS.

Aside from the investigations on natural inhibitors, efforts have been made to identify or design chemical compounds that may target the LuxR-like receptor(s). Most of the designs of inhibitors were based on actual structures of the N-AHSL molecules. These studies led to the identification of analogues with either activating or inhibitory activity (Reverchon et al. 2002; Castang et al. 2004; Smith et al. 2003a). Amongst these latter molecules, phenyl-acyl- and chlorophenyl-acyl-homoserine lactone appear to be the most potent inhibitors. Random screening has also permitted the identification of QS inhibitors such as 4-nitro-pyridine-N-oxide, aniline derivatives, N-methyl-iminocycloheptane, N-methyl-N-iminopyrrolidine and complex heterocycles such as ursolic acid (Smith et al. 2003b; Rasmussen et al. 2005a; Ren et al. 2005).

The N-AHSL signals as targets

QQ may also rely upon signal degradation. N-AHSLs being lactone molecules, they are susceptible to lactonolysis, i.e. the opening of the lactone ring under alkaline pH conditions. The resulting compounds, the cognate N-acyl-homoserines, are not recognized as QS signals by bacteria. Alkaline lactonolysis, a chemical reaction, is subject to the Arrhenius law and is therefore temperature-dependent. Both dependences have been demonstrated in vitro (Byers et al. 2002; Yates et al. 2002; Delalande et al. 2005) and most likely occurs *in planta*. Indeed, several elicitors produced by plant pathogenic bacteria induce a multifaceted plant cell response, one aspect of which is a pH increase (e.g. Bourque et al. 1998). Other signals such as bacterial toxins provoke a similar transient pH increase (Boller 1995) that appears to be a key component of the plant defence systems, related to the expression of defence genes as reported in tomato (Schaller and Oecking 1999).

Most of the QQ studies have dealt, however, with the biological degradation of N-AHSL, first observed in bacteria such as *Variovorax* (Leadbetter and Greenberg 2000) and *Bacillus* (Dong et al. 2000). Since these early reports, numerous bacteria inactivating N-AHSL have been identified. Some dissimilate N-AHSL, i.e. use these substrates as growth substrates; some do not. To date, N-AHSL inactivation has been described in α-proteobacteria, e.g. *Agrobacterium*, *Bosea*, *Sphingopyxis* and *Ochrobactrum* (Zhang et al. 2002; Carlier et al. 2003; D'Angelo-Picard et al. 2005; Jafra et al. 2006), β-proteobacteria, e.g. *Variovorax*, *Ralstonia*, *Comamonas*, and *Delftia* (Leadbetter and Greenberg 2000; Lin et al. 2003; Uroz et al. 2003; Jafra et al. 2006), and γ-proteobacteria, e.g. *Pseudomonas* and *Acinetobacter* (Uroz et al. 2003; Huang et al. 2003; Kang et al. 2004). N-AHSL inactivation also occurs in Gram-positive strains, both amongst low-G + C% strains or firmicutes such as *Bacillus* (Dong et al. 2000, 2002; Fray 2002; Lee et al. 2002; D'Angelo-Picard et al. 2005) and high-G + C% strains or actinobacteria, e.g. *Rhodococcus*, *Arthrobacter*, and *Streptomyces* (Uroz et al. 2003; Park et al. 2003, 2005, 2006).

In bacteria, the N-AHSL-inactivating enzymes described to date belong to two enzymatic families: the N-AHSL lactone hydrolases (e.g. AiiA, AttM and

AiiB; Carlier et al. 2003; Dong et al. 2000, 2002; Lee et al. 2002; Zhang et al. 2002) and the N-AHSL acylases/amidohydrolases (AiiD, PvdQ or AhlM; Lin et al. 2003; Huang et al. 2003; Park et al. 2003, 2005; Uroz et al. 2006). N-AHSL lactone hydrolases catalyse a reaction that is identical to pH-mediated lactonolysis, while acylases/amidohydrolases convert N-AHSL to homoserine lactone and a fatty acid. Lactonases generally hydrolyse a large range of N-AHSLs, from short- (C4- or C6-HSL) to long-chain (C12- and C14-HSL) independently of the substitution at carbon 3 (C3). Though not a systematic phenomenon, amidohydrolase may exhibit a more restricted specificity, being specific for long-chain but not short-chain N-AHSL (Park et al. 2005; Sio et al. 2006). Recently, an N-AHSL-modifying activity has been described in *Rhodococcus erythropolis*; in this species, an oxidoreductase converts 3-oxo,N-AHSL to 3-hydroxy,N-AHSL (Uroz et al. 2005). This activity is not a degradative activity *sensu stricto*; it leads, however, to a change in or loss of the signalling capability of the molecules as the substitution at C3 is crucial for signal specificity.

Aside from bacteria, N-AHSL-degradation abilities have been observed in porcine kidney (Xu et al. 2003) and human airway epithelial cells (Chun et al. 2004). N-AHSL degradation has also been detected in the blood sera of various animals: mouse, rabbit, horse and human sera but remained absent from that of fish and chicken, a possible indication of mammalian specificity (Yang et al. 2005). While porcine kidney cells produce an acylase/amidohydrolase (Xu et al. 2003), N-AHSL degradation by airway epithelial cells is mediated by at least three paraoxonases (PON). One of them exhibits a lactonase activity towards N-AHSL (Draganov et al. 2000; Yang et al. 2005). In plants, degradation of the N-AHSL C6-HSL has been demonstrated in vitro, in the growth medium of seedlings of two legume species: clover and lotus. Under similar conditions, seedlings of corn and wheat did not exhibit any C6-HSL-degradative ability (Delalande et al. 2005).

The biological role of N-AHSL degradation

The biological role of N-AHSL-degrading enzymes, with respect to QS-regulated functions, has been investigated only very recently. One of the most-investigated models is the *attKLM* operon in *Agrobacterium*. In this bacterium, QS controls the conjugal transfer of the Ti plasmid in the presence of opines (for review see: Farrand 1998). The *attKLM* operon is expressed under carbon starvation (Zhang et al. 2004; Wang et al. 2006) or, regardless of the growth phase, in the presence of several plant molecules such as γ-butyrolactone (GBL), γ-hydroxybutyrate (GHB), succinic semialdehyde (SSA), and γ-aminobutyrate (GABA) (Carlier et al. 2004; Chevrot et al. 2006). GBL, GHB, and SSA are dissimilated by the *attKLM*-encoded catabolic pathway, while GABA is not. Remarkably GABA, which would be considered a gratuitous inducer of this operon, is produced at elevated concentrations by wounded plants. Under those conditions, i.e. at the very early stages of infection, the expression of the lactonase gene should therefore be induced. This may permit the degradation of the N-AHSL synthesized by the agrobacteria or other bacteria (Chevrot et al. 2006) and prevent any conjugal transfer from occurring, even though very limited amounts of opines could already be produced by a few transformed cells. The lactonase AttM could therefore participate in the fine regulation of QS in the course of the *Agrobacterium*-plant interaction.

Other biological roles for N-AHSL degradation have been suggested, based on the observation that some N-AHSLs may be toxic to other bacteria. The small bacteriocin produced by several *Rhizobium leguminosarum* strains—which has bacteriostatic activities towards some other strains of this species—is indeed the N-AHSL N-(3R-hydroxy-7-cis-tetradecanoyl)-L-homoserine lactone (Schripsema et al. 1996). The N-AHSL lactonase present in some *Rhizobium* strains may therefore be involved in the inactivation of the small bacteriocin. Similarly, the 3-oxo,dodecanoyl-homoserine lactone (3O,C12-HSL) and its spontaneous reorganization derivative, the tetramic acid 3-(1-hydroxydecylidene)-5-(2-hydroxyethyl)pyrrolidine-2,4-dione, are toxic to several Gram-positive, but not to Gram-negative bacteria (Kaufmann et al. 2005). Both N-AHSL lactonase and acylase/amidohydrolase detected in some firmicutes and actinobacteria may therefore play a key role in detoxifying these compounds.

One cannot exclude the possibility that N-AHSL-cleaving enzymes may be implicated in functions unrelated to QQ, because such enzymes are also capable of degrading other molecules. In *Agrobacterium*, the *attKLM* operon is involved in the dissimilation of GBL, GHB and SSA (Carlier et al. 2004). In *Streptomyces* spp., the cyclic lipopeptide acylase AhlM also degrades penicillin (Park et al. 2005) and possibly related beta-lactame antibiotics. In mammalian cells, N-AHSL paraoxonases were first identified as organophosphate-detoxifying enzymes. They also have thioloactonase activities and exhibit antioxidizing properties, mostly towards sterols and lipid-like molecules (Draganov et al. 2000; Jakubowski 2000).

Though the biological role of QQ is not fully understood, several authors have proposed to take advantage of quenching to develop novel medical and animal therapies (Cámara et al. 2002; Raffa et al. 2005, Rasmussen and Givskov 2006) or novel biocontrol strategies for plant pathogens (Savka et al. 2002; Von Bodman et al. 2003; Zhang 2003). Amongst plant pathogens, those from the *Erwinia/ Pectobacterium* genera have been widely used as model systems to evaluate the validity of the QQ strategy.

Erwinia/Pectobacterium-induced diseases

Overview of *Erwinia/Pectobacterium* taxonomy

The *Erwinia* genus and related genera mostly consist of plant pathogenic bacteria and members of the family *Enterobacteriaceae* (proteobacteria). Like several other microbial groups, the genus *Erwinia* has been reorganized in the light of 16S-based, molecular phylogeny. Four groups have been defined. Group I encompasses species renamed as *Pantoea* species and related bacteria such as *Erwinia herbicola*. Group II includes *Erwinia amylovora* and *Erwinia mallotivora*, group III, species renamed *Pectobacterium*, such as *P. carotovorum* and *P. atrosepticum*, or *Dickeya*, as *D. chrysanthemi*, and group IV consists of strains renamed *Brenneria* but the position of this latter group is still debated (Kwon et al. 1997; Sproer et al. 1999; Gardan et al. 2003; Samson et al. 2005). *Pantoea* species cause bacterial wilt and leaf blight, a disease transmitted by the coleopteran *Chaetocnema pulicaria* (flea beetle). Inoculated to plants, the pathogen produces characteristic water-soaked lesions on young leaves. Eventually, the pathogen colonizes the xylem vessels, leading to subsequent wilting (Von Bodman et al. 2003). Group II *Erwinia* species are the causative agents of bacterial blight, mostly on fruit trees (*E. amylovora*), or bacterial leaf spots (*E. mallotivora*).

Pectobacterium-induced plant diseases

Pectobacterium species are responsible for other disease symptoms known as soft rot, the main symptom being a complete maceration (enzymatic destruction) of plant tissues. Pathogenicity essentially relies upon the production and the secretion by the bacteria of plant-cell-wall-macerating enzymes (PCWME), mostly pectate lyases, pectin methylesterases (which facilitate the action of the first-cited enzymes), pectin lyase, polygalacturonases and oligogalacturonate lyases (for reviews see: Salmond 1994; Hugouvieux-Cotte-Pattat et al. 1996). Other virulence factors of *Pectobacterium* include the production of harpin, a peptide first identified in *Erwinia* in group II *E. amylovora* strains (Wei et al. 1992). Harpin is secreted into plant cells via a type III secretion system (TTSS) encoded by the *hrp* genes (for reviews see: Alfano and Collmer 2004; He 2004). In general, the harpins produced by plant pathogenic bacteria, as well as other TTSS effector peptides such as avirulence gene products, are involved in counter-acting plant defence systems; some of these peptides do exhibit plant-cell-death-inhibiting activity (for review see: Mudgett 2005). In *Pectobacterium*, the precise mode of action of harpin has not been described, though its contribution to pathogenicity has been reported (Bauer et al. 1995; Yang et al. 2002). In addition, harpin could contribute to the aggregative properties of *Pectobacterium* strains (Yap et al. 2006).

Pathogenicity and pathogenicity-related functions in *Pectobacterium* are regulated in a complex manner. Physiologically, environmental parameters affect pathogenicity (Perombelon and Kelman 1980). Pectate lyase synthesis is 20 times higher at 25°C than at 37°C in *D. chrysanthemi* (Hugouvieux-Cotte-Pattat et al. 1992). In several *P. atrosepticum* strains, the emergence of disease symptoms and the expression of maceration enzymes are optimal at temperatures <20°C (Smadja et al. 2004a). Iron deprivation also induces pectate lyase synthesis (Sauvage and Expert 1994), which is also lower under aerobiosis than under reduced oxygen tension, a condition where plant defences are weak. In potato tubers, for instance, *Pectobacterium* may use the naturally-occurring nitrate as a terminal electron acceptor and produce pectate lyases (Hugouvieux-Cotte-Pattat et al. 1992; Smid et al. 1993).

Aside from environmental parameters, pathogenicity in *P. carotovorum* is controlled by at least three factors: (i) the presence of molecules originating from the plant cell wall, (ii) the general GacA/S system, and (iii) QS (for reviews see: Hugouvieux-Cotte-Pattat et al. 1996; Whitehead et al. 2001, 2002; Von Bodman et al. 2003; Fig. 1). To summarize and simplify, in *P. carotovorum*, in the plant environment and at low cell density, only limited amounts of 3-oxo,hexanoyl-homoserine lactone (3O,C6-HSL) are synthesized by the CarI N-AHSL synthase (Jones et al. 1993). Under those conditions, the expression of the genes encoding plant-cell-wall-maceration enzymes (PCWME) is blocked at: (i) the transcriptional level, a phenomenon mediated by the KdgR repressor (Nasser et al. 1994), and (ii) the post-transcriptional level, a feature mediated by the RsmA protein that binds the PCWME mRNA and accelerates its degradation (Cui et al. 1995). As a consequence, very limited amounts of the PCWME are produced. At high cell density, the presence of an elevated concentration of 3O,C6-HSL is sensed by the regulatory protein ExpR that, as an ExpR/N-AHSL complex, prevents the transcription of *rsmA* which, conversely, is activated in the absence of the cognate N-AHSL (Cui et al. 2005). The existing RsmA protein is further displaced from the PCWME mRNA by the activation of the production of RsmB, a small RsmA-binding RNA, encoded by the eponymous gene *rsmB* (Liu et al. 1998), the transcription of which is controlled by the global regulatory system ExpS/ExpA (analogous to GacS/GacA; Cui et al. 2001; Hyytiainen et al. 2001). PCWME are therefore synthesized, leading to the degradation of plant pectin. Oligomers and degradation products generated by the enzymes, such as polygalacturonate, saturated and unsaturated digalacturonate, galacturonate, and mostly 2-keto, 3-deoxygluconate, recognized by the repressor KdgR (Nasser et al. 1994), further induce the expression of both *rsmB* and PCWME genes, leading to an extensive production of PCWME (Tsuyumu 1977; Collmer and Bateman 1981).

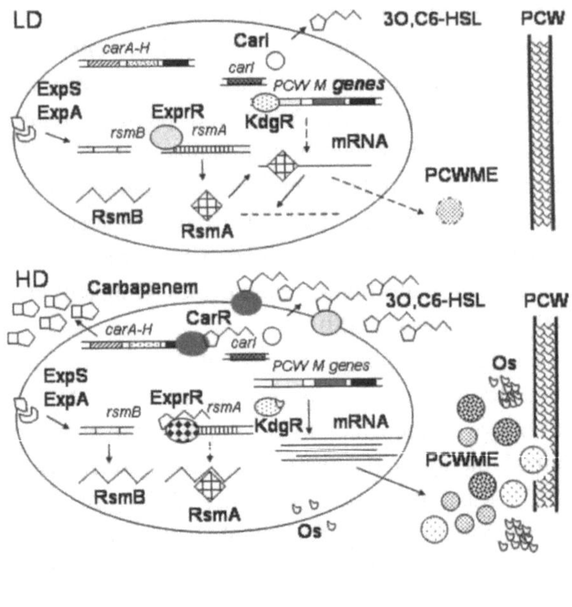

Fig. 1 Regulation of plant-cell-wall-macerating enzymes in *Pectobacterium*. At low cell density (panel LD, top), limited amounts of the N-AHSL 3O,C6-HSL are made via the CarI synthase. Transcription of the genes involved in plant-cell-wall maceration (PCWM) is blocked both by the repressor KdgR and by the RNA-binding protein RsmA. The plant-cell-wall-maceration enzymes (PCWME) are not produced. At high cell density (panel HD, bottom), large amounts of 3O,C6-HSL are made, preventing the synthesis of RsmA via the ExpR-N-AHSL complex. The presence in the cell of the RsmA-binding RNA RsmB, the production of which depends upon the global ExpS/A regulatory system, allows further trapping of RsmA. The PCWM genes are partly expressed leading to the production of PCWME. The enzymatic activities generate oligosides (Os) from the cell wall (PCW). Os release the KdgR-mediated repression of RsmB and PCWM genes. The QS signals also allow the activation of the regulator CarR that controls the expression of the *carAH* genes involved in the synthesis of carbapenem. For additional details, see text

The production of harpin encoded by the *hrpN* gene in *P. carotovorum* is controlled by the same three factors that regulate the synthesis of PCWME, i.e. (i) the presence of molecules originating from the plant cell wall (Liu et al. 1999), (ii) the general GacA/S system (Cui et al. 2001), and (iii) QS via the RsmA/B system (Cui et al. 1995). In addition, harpin production is also modulated by environmental parameters, a response mediated by the product of the *hrpL* regulatory gene (Wei and Beer 1995), which—by analogy with peptides encoded by *hrpL* orthologues—is most likely an alternate sigma factor (Chaterjee et al. 2002). However, the precise regulatory mechanism for hairpin production in *Pectobacterium* remains only partly understood.

Aside from pathogenicity, another function is controlled by QS in *Pectobacterium*: the production of the lactame antibiotic carbapenem (for reviews see: Whitehead et al. 2001, 2002; Von Bodman et al. 2003). When accumulated at a sufficient concentration, the signal 3O,C6-HSL produced by the synthase CarI is perceived by the regulator CarR which activates the transcription of the carA-K genes, encoding the enzymes involved in the biosynthesis of the antibiotic (McGowan et al. 1995, 1996). Carbapenem production most likely confers a fitness advantage to *Pectobacterium*, possibly by reducing the number of competing bacteria (Whitehead et al. 2001, 2002) in soil or in plants, macerated or not. However, resistance to β-lactame in bacteria living in the plant environment is widespread (Ogawara 1981). It is therefore possible that another advantage linked to carbapenem production lies in the resistance of *Pectobacterium* to carbapenem and related antibiotics produced by competing bacteria.

QQ of *Pectobacterium* pathogenicity

The above description of the molecular mechanisms underlying pathogenicity in *Pectobacterium* highlights the central role played by QS regulation in pathogenicity. Targeting the QS regulatory elements to develop biocontrol strategies for *Pectobacterium* is therefore a pertinent option (Dong et al. 2000; Smadja et al. 2004b). Two research strategies have been developed: one aimed at producing transgenic plants interfering with QS, the other at isolating plant-associated bacteria naturally interfering with QS in *Pectobacterium.*

The plant-genetic-engineering approach

Plants were genetically modified to gain the capacity to produce or inactivate N-AHSL signals. A first series of these transgenic plants were developed to activate QS functions of pathogens at an inappropriate time; a second type of plants was designed to block the initiation of the QS regulatory cascade. Transgenic tobacco plants, into which the *yenI* gene of *Yersinia enterolitica* encoding N-AHSL synthase was introduced, were able to produce C6-HSL and 3O,C6-HSL (Fray et al. 1999). The N-AHSL-producing plants was able to complement the virulence of an N-AHSL-defective mutant of *P. carotovorum*, as well as the biocontrol activity of an N-AHSL-defective mutant of *Pseudomonas aureofaciens*. However, while a decrease in the virulence of a wild-type *P. carotovorum* strain on the non-host tobacco plant expressing *expI* was reported (Mäe et al. 2001), an increase in the virulence was observed when wild-type *P. carotovorum* was inoculated on the host potato plant expressing the *yenI* gene (Toth et al. 2004). Quite different results were obtained with transgenic tobacco and potato plants expressing the lactonase AiiA of *Bacillus* sp. 240B1 (Dong et al. 2001). These *aiiA*-plants, expressing lactonase activity directed at N-AHSL, were always more resistant to *P. carotovorum* infection than the parental, wild-type plants.

The biocontrol approach

In consideration of the debate that exists in Europe on the use and release of GM plants (e.g. Hodgson 2001; Williams 2002; Wisniewski et al. 2002), a more acceptable biocontrol approach was developed by various researchers. Several studies aimed at isolating bacteria able to inactivate the N-AHSL signals produced by *Pectobacterium.* These studies have been facilitated by the occurrence of N-AHSL-degrading bacteria in soil and plant environments. This community represents 2–5 and up to 10% of the culturable bacteria (Dong et al. 2000; Steidle et al. 2001; Morello et al. 2004; D'Angelo-Picard et al. 2004, 2005), a feature that translates into a demonstrable N-AHSL-degradation potential for soils (Wang and Leadbetter 2005).

Bacterial populations from bare and rhizospheric soil could be screened for N-AHSL degraders by randomly assaying the N-AHSL-inactivation capability of individual isolates in vitro. Using this strategy, *Bacillus* strains exhibiting the AiiA-borne lactonase activity were identified (Dong et al. 2000). A similar experimental design led to the identification of a *Ralstonia* strain from a complex biofilm population (Lin et al. 2003). A mass screen of bacteria isolated from the root system of wild-type plants and plants producing N-AHSL permitted the identification of additional *Bacillus* strains, *Agrobacterium* spp., *Sphingopyxis witflariensis* and *Bosea thiooxi-*

dans isolates inactivating N-AHSL (D'Angelo-Picard et al. 2005). However, none of these isolates have been used against *Pectobacterium* strains in biocontrol experiments. The strategy proved to be valuable, however, as it allowed the isolation of an *Acinetobacter* strain which degraded N-AHSL and was capable of attenuating soft-rot symptoms caused by *P. carotovorum* in potato tuber slice assays (Kang et al. 2004).

Another strategy, close to but distinct from the one described above, aimed at isolating bacteria with N-AHSL-dissimilating ability by selection on minimal media supplemented with N-AHSL as the sole carbon source. The prototypic experiment led to the identification of the first degrader, a *Variovorax paradoxus* strain (Leadbetter and Greenberg 2000), and later to the demonstration of the ability of the PAO1 strain of *Pseudomonas aeruginosa* to dissimilate long chain N-AHSLs such as 3O,C12-HSL (Huang et al. 2003). Valuable biocontrol strains (also termed quenchers) directed against pathogenic *Pectobacterium* were also isolated using this technique; examples include *Comamonas* spp., *Ochrobactrum*, and *Rhodococcus* strains (Uroz et al. 2003, 2005, 2006; Jaffra et al. 2006; Park et al. 2006). The remarkable ability of several *Rhodococcus* strains to quench pathogenicity in *Pectobacterium* (complete disappearance of disease symptoms at 1 to 1 and 1 to 10 ratios), though variable as a function of the origin of the strain, possibly is due to the occurrence, in these bacteria, of a triple inactivation pathway consisting of an acyclase/amidohydrolase, a lactonase and an oxidoreductase (Uroz et al. 2005; Park et al. 2006).

Perspectives in QQ

Targeting the QS-regulated virulence functions of *Pectobacterium* should not cause the disappearance of the pathogen from the plant environment, since these functions are not vital to *Pectobacterium*. This circumstance could lead to the appearance of healthy, contaminated plants, from which disease may spread in the absence of the quencher treatment. Also, QQ strategies being non-selective for the time being, they may also prevent bacterial functions possibly beneficial to plants (such as antifungal synthesis) (Molina et al. 2003). Whether these points constitute major drawbacks in the development of QQ strategies aimed at *Pectobacterium* remains to be evaluated.

At this time, none of the isolated quenchers have been assayed for their biocontrol ability outside the laboratory, a step which remains crucial (and often time-consuming) to evaluate the potential value of a biocontrol agent under agricultural conditions (Mc Intyre and Press 1991). However, the commercial use of transgenic plants expressing N-AHSL-degrading enzymes essentially depends on the legal authorization given by each State.

In addition to lactonases of the AiiA family and amidohydrolases of the AiiD family, other genes could be used if they confer an increased N-AHSL-degradation ability upon the host plants. Known genes such as those encoding paraoxonase (Yang et al. 2005) may be candidates, along with genes originating from unculturable bacteria that represent the vast majority of bacterial soil inhabitants (e.g. Felkse et al. 1999; for review see: Saleh-Lakha et al. 2005). Such genes have been detected in soil bacteria via a metagenomic approach (Williamson et al. 2005; our laboratory, unpublished) but the mechanism that led to N-AHSL degradation or inhibition of N-AHSL detection remains unknown. Finally, the increased number of QQ chemicals offers a third strategy to neutralize QS pathogens. The biocontrol, transgenic and chemical approaches are not exclusive, however, and represent complementary ways to fight QS-regulated virulence in *Pectobacterium* and related plant pathogens.

References

Alfano, J. R., & Collmer, A. (2004). Type III secretion system effector proteins: double agents in bacterial disease and plant defense. *Annual Review of Phytopathology, 42*, 385–414.

Bauer, D. W., Wei, Z. M., Beer, S. V., & Collmer, A. (1995). *Erwinia chrysanthemi* harpinEch: An elicitor of the hypersensitive response that contributes to soft-rot pathogenesis. *Molecular Plant-Microbe Interactions, 8*, 484–491.

Boller, T. (1995). Chemoperception of microbial signals in plant cells. *Annual Review of Plant Physiology and Plant Molecular Biology, 46*, 189–214.

Bourque, S., Ponchet, M., Binet, M.-N., Ricci, P., Pugin, A., & Lebrun-Garcia, A. (1998). Comparison of binding properties and early biological effects of elicitins in tobacco cells. *Plant Physiology, 118*, 1317–1326.

Byers, J. T., Lucas, C., Salmond, G. P., & Welch, M. (2002). Nonenzymatic turnover of an *Erwinia carotovora* quorum-sensing signaling molecule. *Journal of Bacteriology, 184*, 1163–1171.

Cámara, M., Williams, P., & Hardman, A. (2002). Controlling infection by tuning in and turning down the volume of bacterial small-talk. *The Lancet Infectious Diseases, 2*, 667–676.

Carlier, A., Chevrot, R., Dessaux, Y., & Faure, D. (2004). The assimilation of gamma-butyrolactone in *Agrobacterium tumefaciens* C58 interferes with the accumulation of the N-acyl-homoserine lactone signal. *Molecular Plant-Microbe Interactions, 17*, 951–957.

Carlier, A., Uroz, S., Smadja, B., Fray, R., Latour, X., Dessaux, Y., & Faure, D. (2003). The Ti plasmid of *Agrobacterium tumefaciens* harbors an *attM* paralogous gene *aiiB* also encoding N-acylhomoserine lactonase activity. *Applied and Environmental Microbiology, 69*, 4989–4993.

Castang, S., Chantegrel, B., Deshayes, C., Dolmazon, R., Gouet, P., Haser, R., Reverchon, S., Nasser, W., Hugouvieux-Cotte-Pattat, N., & Doutheau, A. (2004). N-Sulfonyl homoserine lactones as antagonists of bacterial quorum sensing. *Bioorganic & Medicinal Chemistry Letters, 14*, 5145–5149.

Chatterjee, A., Cui, Y., & Chatterjee, A. K. (2002). Regulation of *Erwinia carotovora hrpL* (sigma-L), which encodes an extracytoplasmid function subfamily of sigma factor required for the expression of the HRP regulon. *Molecular Plant-Microbe Interactions, 15*, 971–980.

Che, X., Schauder, S., Potier, N., Van Dorsselaer, A., Pelczer, I., Bassler, B. L., & Hughson, F. M. (2002). Structural identification of a bacterial quorum-sensing signal containing boron. *Nature, 415*, 545–549.

Chevrot, R., Rosen, R., Haudecoeur, E., Cirou, A., Shelp, B. J., Ron, E., & Faure, D. (2006). GABA controls the level of quorum-sensing signal in *Agrobacterium tumefaciens. Proceedings of the National Academy of Sciences of the USA, 103*, 7460–7464.

Choo, J. H., Rukayadi, Y., & Hwang, J. K. (2006). Inhibition of bacterial quorum sensing by vanilla extract. *Letters in Applied Microbiology, 42*, 637–641.

Chun, C. K., Ozer, E. A., Welsh, M. J., Zabner, J., & Greenberg, E. P. (2004). Inactivation of a *Pseudomonas aeruginosa* quorum-sensing signal by human airway epithelia. *Proceedings of the National Academy of Sciences of the USA, 101*, 3587–3590.

Colin Slaughter, J. (1999). The naturally occurring furanones: Formation and function from pheromone to food. *Biological Reviews of the Cambridge Philosophical Society, 74*, 259–276.

Collmer, A., & Bateman, D. F. (1981). Impaired induction and self-catabolite repression of extracellular pectate lyase in *Erwinia chrysanthemi* mutants deficient in oligogalacturonide lyase. *Proceedings of the National Academy of Sciences of the USA, 78*, 3920–3924.

Cui, Y., Chatterjee, A., & Chatterjee, A. K. (2001). Effects of the two-component system comprising GacA and GacS of *Erwinia carotovora* subsp. *carotovora* on the production of global regulatory rsmB RNA, extracellular enzymes, and harpinEcc. *Molecular Plant-Microbe Interactions, 14*, 516–526.

Cui, Y., Chatterjee, A., Hasegawa, H., Dixit, V., Leigh, N., & Chatterjee, A. K. (2005). ExpR, a LuxR homolog of *Erwinia carotovora* subsp. *carotovora*, activates transcription of *rsmA*, which specifies a global regulatory RNA-binding protein. *Journal of Bacteriology, 187*, 4792–4803.

Cui, Y., Chatterjee, A., Liu, Y., Dumenyo, C. K., & Chatterjee, A. K. (1995). Identification of a global repressor gene, *rsmA*, of *Erwinia carotovora* subsp. *carotovora* that controls extracellular enzymes, N-(3-oxohexanoyl)-L-homoserine lactone, and pathogenicity in soft-rotting *Erwinia* spp. *Journal of Bacteriology, 177*, 5108–5115.

D'Angelo-Picard, C., Faure, D., Carlier, A., Uroz, S., Raffoux, A., Fray, R., & Dessaux, Y. (2004). Bacterial populations in the rhizosphere of tobacco plants producing the quorum-sensing signals hexanoyl-homoserine lactone and 3-oxo-hexanoyl-homoserine lactone. *FEMS Microbiology Ecology, 51*, 19–29.

D'Angelo-Picard, C., Faure, D., Penot, I., & Dessaux, Y. (2005). Diversity of N-acyl homoserine lactone-producing and -degrading bacteria in soil and tobacco rhizosphere. *Environmental Microbiology, 7*, 1796–1808.

Delalande, L., Faure, D., Raffoux, A., Uroz, S., D'Angelo, C., Elasri, M., Carlier, A., Berruyer, R., Petit, A., Williams, P., & Dessaux, Y. (2005). N-Hexanoyl-L-homoserine lactone, a mediator of bacterial quorum-sensing regulation, exhibits a plant-dependent stability in the rhizosphere and may be inactivated by germinating *Lotus corniculatus* seedlings. *FEMS Microbiology Ecology, 52*, 13–20.

Dong, Y. H., Gusti, A. R., Zhang, Q., Xu, J. L., & Zhang, L. H. (2002). Identification of quorum-quenching N-acyl homoserine lactonases from *Bacillus* species. *Applied and Environmental Microbiology, 68*, 1754–1759.

Dong, Y. H., Wang, L. H., Xu, J. L., Zhang, H. B., Zhang, X. F., & Zhang, L. H. (2001). Quenching quorum-sensing-dependent bacterial infection by an N-acyl homoserine lactonase. *Nature, 411*, 813–817.

Dong, Y. H., Xu, J. L., Li, X. Z., & Zhang, L. H. (2000). AiiA, an enzyme that inactivates the acylhomoserine lactone quorum-sensing signal and attenuates the virulence of *Erwinia carotovora. Proceedings of the National Academy of Sciences of the USA, 97*, 3526–3531.

Dong, Y. H., & Zhang, L. H. (2005). Quorum sensing and quorum-quenching enzymes. *Journal of Microbiology, 43*, 101–109.

Draganov, D. I., Stetson, P. L., Watson, C. E., Billecke, S. S., & La Du, B. N. (2000). Rabbit Serum Paraoxonase 3 (PON3) is a high density lipoprotein-associated lactonase and protects low density lipoprotein against oxidation. *The Journal of Biological Chemistry, 275*, 33435–33442.

Farrand, S. K. (1998). Conjugal plasmids and their transfer. In H. P. Spaink, A. Kondorosi, & P. J. J. Hooykaas (Eds.), *The Rhizobiaceae* (pp. 199–223). Dordrecht (NL): Kluwer Acad. Pub.

Felske, A., Wolterink, A., van Lis, R., de Vos, W. M., & Akkermans, A. D. L. (1999). Searching for the predominant soil bacteria: 16S rDNA cloning versus strain cultivation. *FEMS Microbiology Ecology, 30*, 137–145.

Flavier, A. B., Clough, S. J., Schell, M. A., & Denny, T. P. (1997). Identification of 3-hydroxypalmitic acid methyl ester as a novel autoregulator controlling virulence in *Ralstonia solanacearum. Molecular Microbiology, 26*, 251–259.

Frank, H. K. (1977). Occurrence of patulin in fruit and vegetables. *Annales de la nutrition et de l'alimentation, 31*, 459–465.

Fray, R. G. (2002). Altering plant-microbe interaction through artificially manipulating bacterial quorum sensing. *Annals of Botany (London), 89*, 245–253.

Fray, R. G., Throup, J. P., Daykin, M., Wallace, A., Williams, P., Stewart, G. S., & Grierson, D. (1999). Plants genetically modified to produce N-acylhomoserine lactones communicate with bacteria. *Nature Biotechnology, 17*, 1017–1020.

Fuqua, C., & Greenberg, E. P. (1998). Self perception in bacteria: Quorum sensing with acylated homoserine lactones. *Current Opinion in Microbiology, 1*, 183–189.

Fuqua, C., & Greenberg, E. P. (2002). Listening in on bacteria: Acyl-homoserine lactone signalling. *Nature Reviews Molecular Cell Biology, 3*, 685–695.

Fuqua, C., Parsek, M. R., & Greenberg, E. P. (2001). Regulation of gene expression by cell-to-cell communication: Acyl-homoserine lactone quorum sensing. *Annual Review of Genetics, 35*, 439–468.

Fuqua, W. C., Winans, S. C., & Greenberg, E. P. (1994). Quorum sensing in bacteria: The LuxR/LuxI family of cell density-responsive transcriptional regulators. *Journal of Bacteriology, 176*, 269–275.

Gao, M., Teplitski, M., Robinson, J. B., & Bauer, W. D. (2003). Production of substances by *Medicago truncatula* that affect bacterial quorum sensing. *Molecular Plant-Microbe Interactions, 16*, 827–834.

Gardan, L., Gouy, C., Christen, R., & Samson, R. (2003). Elevation of three subspecies of *Pectobacterium carotovorum* to species level: *Pectobacterium atrosepticum* sp. nov., *Pectobacterium betavasculorum* sp. nov. and *Pectobacterium wasabiae* sp. nov. *International Journal of Systematic and Evolutionary Microbiology, 53*, 381–391.

Geze, M., Blanchard, P., Fourrey, J. L., & Robert-Gero, M. (1983). Synthesis of sinefungin and its C-6′ epimer. *Journal of the American Chemical Society, 105*, 7638–7640.

Givskov, M., de Nys, R., Manefield, M., Gram, L., Maximilien, R., Eberl, L., Molin, S., Steinberg, P. D., & Kjelleberg, S. (1996). Eukaryotic interference with homoserine lactone-mediated prokaryotic signalling. *Journal of Bacteriology, 178*, 6618–6622.

Greenberg, E. P. (2000). Acyl-homoserine lactone quorum sensing in bacteria. *Journal of Microbiology,* 38, 117–121.

He, S. Y. (2004). Type III protein secretion mechanism in mammalian and plant pathogens. *Biochimica et Biophysica Acta, 1694*, 181–206.

Hoang, T. T., & Schweizer, H. P. (1999). Characterization of *Pseudomonas aeruginosa* enoyl-acyl carrier protein reductase (FabI): A target for the antimicrobial triclosan and its role in acylated homoserine lactone synthesis. *Journal of Bacteriology, 181*, 5489–5497.

Hodgson, J. (2001). FoE urges UK to disclose more of GM crop trial data. *Nature Biotechnology, 19*, 699–700.

Holden, M. T., Ram Chhabra, S., de Nys, R., Stead, P., Bainton, N. J., et al. (1999). Quorum-sensing cross talk: Isolation and chemical characterization of cyclic dipeptides from *Pseudomonas aeruginosa* and other gram-negative bacteria. *Molecular Microbiology, 33*, 1254–66.

Huang, J. J., Han, J. I., Zhang, L. Z., & Leadbetter, J. R. (2003). Utilization of acyl-homoserine lactone quorum signals for growth by a soil pseudomonad and *Pseudomonas aeruginosa* PAO1. *Applied and Environmental Microbiology, 69*, 5941–5949.

Hugouvieux-Cotte-Pattat, N., Condemine, G., Nasser, W., & Reverchon, S. (1996). Regulation of pectinolysis in *Erwinia chrysanthemi*. *Annual Review of Microbiology, 50*, 213–257.

Hugouvieux-Cotte-Pattat, N., Dominguez, H., & Robert-Baudouy, J. (1992). Environmental conditions affect the transcription of the pectinase genes of *Erwinia chrysanthemi* 3937. *Journal of Bacteriology, 174*, 7807–7818.

Hyytiainen, H., Montesano, M., & Palva, E. T. (2001). Global regulators ExpA (GacA) and KdgR modulate extracellular enzyme gene expression through the RsmA-rsmB system in *Erwinia carotovora* subsp. *carotovora*. *Molecular Plant-Microbe Interactions, 14*, 931–938.

Jafra, S., Przysowa, J., Czajkowski, R., Michta, A., Garbeva, P., & Van der Wolf, J. M. (2006). Detection and characterization of bacteria from the potato rhizosphere degrading N-acyl-homoserine lactone. *Canadian Journal of Microbiology, 52*, 1006–1015.

Jakubowski, H. (2000). Calcium-dependent human serum homocysteine thiolactone hydrolase. A protective mechanism against protein *N*-homocysteinylation. *The Journal of Biological Chemistry, 275*, 3957–3962.

Jones, S., Yu, B., Bainton, N. J., Birdsall, M., Bycroft, B. W., Chhabra, S. R., Cox, A. J., Golby, P., Reeves, P. J., Stephens, S., Winson, M. K., Salmond, G. P. C., Stewart, G. S. A. B., & Williams, P. (1993). The lux autoinducer regulates the production of exoenzyme virulence determinants in *Erwinia carotovora* and *Pseudomonas aeruginosa*. *EMBO Journal, 12*, 2477–2482.

Kang, B. R., Lee, J. H., Ko, S. J., Lee, Y. H., Cha, J. S., Cho, B. H., & Kim, Y. C. (2004). Degradation of acyl-homoserine lactone molecules by *Acinetobacter* sp strain C1010. *Canadian Journal of Microbiology, 50*, 935–941.

Karamanoli, K., & Lindow, S. E. (2006). Disruption of N-acyl homoserine lactone-mediated cell signaling and iron acquisition in epiphytic bacteria by leaf surface compounds. *Applied Environmental Microbiology, 72*, 7678–7686.

Kaufmann, G. F., Sartorio, R., Lee, S. H., Rogers, C. J., Meijler, M. M., Moss, J. A., Clapham, B., Brogan, A. P., Dickerson, T. J., & Janda, K. D. (2005). Revisiting quorum sensing: Discovery of additional chemical and biological functions for 3-oxo-N-acylhomoserine lactones. *Proceedings of the National Academy of Sciences of the USA, 102*, 309–314.

Kwon, S. W., Go, S. J., Kang, H. W., Ryu, J. C., & Jo, J. K. (1997). Phylogenetic analysis of *Erwinia* species based on 16S rRNA gene sequences. *International Journal of Systematic Bacteriology, 47*, 1061–1067.

Leadbetter, J. R., & Greenberg, E. P. (2000). Metabolism of acyl-homoserine lactone quorum-sensing signals by *Variovorax paradoxus*. *Journal of Bacteriology, 182*, 6921–6926.

Lee, S. J., Park, S. Y., Lee, J. J., Yum, D. Y., Koo, B. T., & Lee, J. K. (2002). Genes encoding the N-acyl homoserine lactone-degrading enzyme are widespread in many subspecies of *Bacillus thuringiensis*. *Applied and Environmental Microbiology, 68*, 3919–3924.

Lin, Y. H., Xu, J. L., Hu, J., Wang, L. H., Ong, S. L., Leadbetter, J. R., & Zhang, L. H. (2003). Acyl-homoserine lactone acylase from *Ralstonia* strain XJ12B represents a novel and potent class of quorum-quenching enzymes. *Molecular Microbiology, 47*, 849–860.

Liu, Y., Cui, Y., Mukherjee, A., & Chatterjee, A. K. (1998). Characterization of a novel RNA regulator of *Erwinia carotovora* ssp. *carotovora* that controls production of extracellular enzymes and secondary metabolites. *Molecular Microbiology, 29*, 219–234.

Liu, Y., Jiang, G., Cui, Y., Mukherjee, A., Ma, W. L., & Chatterjee, A. K. (1999). $kdgR_{Ecc}$ negatively regulates genes for pectinases, cellulase, protease, $Harpin_{Ecc}$, and a global RNA regulator in *Erwinia carotovora* subsp. *carotovora*. *Journal of Bacteriology, 181*, 2411–2421.

Mäe, A., Montesano, M., Koiv, V., & Palva, E. T. (2001). Transgenic plants producing the bacterial pheromone N-acyl-homoserine lactone exhibit enhanced resistance to the bacterial phytopathogen *Erwinia carotovora*. *Molecular Plant-Microbe Interactions, 14*, 1035–1042.

Manefield, M., de Nys, R., Kumar, N., Read, R., Givskov, M., Steinberg, P., & Kjelleberg, S. (1999). Evidence that halogenated furanones from *Delisea pulchra* inhibit acylated homoserine lactone (AHL)-mediated gene expression by displacing the AHL signal from its receptor protein. *Microbiology, 145*, 283–291.

Manefield, M., Rasmussen, T. B., Henzter, M., Andersen, J. B., Steinberg, P., Kjelleberg, S., & Givskov, M. (2002). Halogenated furanones inhibit quorum sensing through accelerated LuxR turnover. *Microbiology, 148*, 1119–1127.

McGowan, S., Sebaihia, M., Jones, S., Yu, B., Bainton, N., et al. (1995). Carbapenem antibiotic production in *Erwinia carotovora* is regulated by CarR, a homologue of the LuxR transcriptional activator. *Microbiology, 141*, 541–50.

McGowan, S. J., Sebaihia, M., Porter, L. E., Stewart, G. S., Williams, P., et al. (1996). Analysis of bacterial carbapenem antibiotic production genes reveals a novel ß-lactam biosynthesis pathway. *Molecular Microbiology, 22*, 415–426.

McIntyre, J. L., & Press, L. S. (1991). Formulation, delivery system and marketing of biocontrol agents and plant growth promoting rhizobacteria (PGPR). In D. L. Keister & P. B. Cregan (Eds.), *The rhizosphere and plant growth* (pp. 289–295). Dordrecht (NL): Kluwer Acad Pub.

Molina, L., Constantinescu, F., Michel, L., Reimmann, C., Duffy, B., & Défago, G. (2003). Degradation of pathogen quorum-sensing molecules by soil bacteria: A preventive and curative biological control mechanism. *FEMS Microbiology Ecology, 45*, 71–81.

Morello, J. E., Pierson, E. A., & Pierson, L. S. III (2004). Negative cross-communication among wheat rhizosphere bacteria: Effect on antibiotic production by the biological control bacterium *Pseudomonas aureofaciens* 30–84. *Applied and Environmental Microbiology, 70*, 3103–3109.

Mudgett, M. B. (2005). New insights to the function of phytopathogenic bacterial type III effectors in plants. *Annual Review of Plant Biology, 56*, 509–531.

Nasser, W., Reverchon, S., Condemine, G., & Robert-Baudouy, J. (1994). Specific interactions of *Erwinia chrysanthemi* KdgR repressor with different operators of genes involved in pectinolysis. *Journal of Molecular Biology, 236*, 427–440.

Ogawara, H. (1981). Antibiotic resistance in pathogenic and producing bacteria, with special reference to beta-lactam antibiotics. *Microbiological Reviews, 45*, 591–619.

Park, S. Y., Hwang, B. J., Shin, M. H., Kim, J. A., Kim, H. K., & Lee, J. K. (2006). N-acylhomoserine lactonase producing *Rhodococcus* spp. with different AHL-degrading activities. *FEMS Microbiology Letters, 261*, 102–108.

Park, S. Y., Lee, S. J., Oh, T. K., Oh, J. W., Koo, B. T., Yum, D. Y., & Lee, J. K. (2003). AhlD an N-acylhomoserine lactonase in *Arthrobacter* sp and predicted homologues in other bacteria. *Microbiology, 149*, 1541–1550.

Park, S. Y., Kang, H. O., Jang, H. S., Lee, J. K., Koo, B. T., & Yum, D. Y. (2005). Identification of extracellular N-Acylhomoserine lactone acylase from a *Streptomyces* sp and its application to quorum quenching. *Applied Environmental Microbiology, 71*, 2632–2641.

Parsek, M. R., Val, D. L., Hanzelka, B. L., Cronan, J. E., & Greenberg, E. P. (1999). Acyl homoserine-lactone quorum-sensing signal generation. *Proceedings of the National Academy of Sciences of the USA, 96*, 4360–4365.

Perombelon, M. C. M., & Kelman, A. (1980). Ecology of the soft rot erwinias. *Annual Review of Phytopathology, 18*, 361–387.

Raffa, R. B., Iannuzzo, J. R., Levine, D. R., Saeid, K. K., Schwartz, R. C., Sucic, N. T., Terleckyj, O. D., & Young, J. M. (2005). Bacterial communication ("quorum sensing") via ligands and receptors: A novel pharmacologic target for the design of antibiotic drugs. *The Journal of Pharmacology and Experimental Therapeutics, 312*, 417–423.

Rasmussen, T. B., Bjarnsholt, T., Skindersoe, M. E., Hentzer, M., Kristoffersen, P., Kote, M., Nielsen, J., Eberl, L., & Givskov, M. (2005a). Screening for quorum-sensing inhibitors (QSI) by use of a novel genetic system, the QSI selector. *Journal of Bacteriology, 187*, 1799–1814.

Rasmussen, T. B., & Givskov, M. (2006). Quorum sensing inhibitors: A bargain of effects. *Microbiology, 152*, 895–904.

Rasmussen, T. B., Manefield, M., Andersen, J. B., Eberl, L., Anthoni, U., Christophersen, C., Steinberg, P., Kjelleberg, S., & Givskov, M. (2000). How *Delisea pulchra* furanones affect quorum sensing and swarming motility in *Serratia liquefaciens* MG1. *Microbiology, 146*, 3237–3244.

Rasmussen, T. B., Skindersoe, M. E., Bjarnsholt, T., Phipps, R. K., Christensen, K. B., Jensen, P. O., Andersen, J. B., Koch, B., Larsen, T. O., Hentzer, M., Eberl, L., Hoiby, N., & Givskov, M. (2005b). Identity and effects of quorum-sensing inhibitors produced by *Penicillium* species. *Microbiology, 151*, 1325–1340.

Redfield, R. J. (2002). Is quorum sensing a side effect of diffusion sensing? *Trends Microbiology, 10*, 365–370.

Ren, D., Zuo, R., Gonzalez Barrios, A. F., Bedzyk, L. A., Eldridge, G. R., Pasmore, M. E., & Wood, T. K. (2005). Differential gene expression for investigation of *Escherichia coli* biofilm inhibition by plant extract ursolic acid. *Applied and Environmental Microbiology, 71*, 4022–4034.

Reverchon, S., Chantegrel, B., Deshayes, C., Doutheau, A., & Cotte-Pattat, N. (2002). New synthetic analogues of N-acyl homoserine lactones as agonists or antagonists of transcriptional regulators involved in bacterial quorum sensing. *Bioorganic & Medicinal Chemistry Letters, 12*, 1153–1157.

Saleh-Lakha, S., Miller, M., Campbell, R. G., Schneider, K., Elahimanesh, P., Hart, M. M., & Trevors, J. T. (2005). Microbial gene expression in soil: Methods, applications and challenges. *Journal of Microbiology Methods, 63*, 1–19.

Salmond, G. P. C. (1994). Secretion of extracellular virulence factors by plant pathogenic bacteria. *Annual Review of Phytopathology, 32*, 181–200.

Samson, R., Legendre, J. B., Christen, R., Fischer-Le Saux, M., Achouak, W., & Gardan, L. (2005). Transfer of *Pectobacterium chrysanthemi* (Burkholder et al. 1953) Brenner et al. 1973 and *Brenneria paradisiaca* to the genus *Dickeya gen.* nov. as *Dickeya chrysanthemi* comb. nov. and *Dickeya paradisiaca* comb. nov. and delineation of four novel species, *Dickeya dadantii* sp. nov., *Dickeya dianthicola* sp. nov., *Dickeya dieffenbachiae* sp. nov. and *Dickeya zeae* sp. nov. *International Journal of Systematic and Evolutionary Microbiology, 55*, 1415–1427.

Sauvage, C., & Expert, D. (1994). Differential regulation by iron of *Erwinia chrysanthemi* pectate lyases: Pathogenicity of iron transport regulatory (*cbr*) mutants. *Molecular Plant-Microbe Interactions, 7*, 71–77.

Savka, M. A., Dessaux, Y., Oger, P., & Rossbach, S. (2002). Engineering bacterial competitiveness and persistence in the phytosphere. *Molecular Plant-Microbe Interactions, 15*, 866–874.

Schaller, A., & Oecking, C. (1999). Modulation of plasma membrane H^+-ATPase activity differentially activates wound and pathogen defense responses in tomato plants. *Plant Cell, 11*, 263–272.

Schripsema, J., de Rudder, K. E., van Vliet, T. B., Lankhorst, P. P., de Vroom, E., Kijne, J. W., & van Brussel, A. A. (1996). Bacteriocin small of *Rhizobium leguminosarum* belongs to the class of N-acyl-L-homoserine lactone molecules, known as autoinducers and as quorum sensing co-transcription factors. *Journal of Bacteriology, 178*, 366–371.

Scott, P. M., Miles, W. F., Toft, P., & Dube, J. G. (1972). Occurrence of patulin in apple juice. *Journal of Agricultural and Food Chemistry, 20*, 450–451.

Sio, C. F., Otten, L. G., Cool, R. H., Diggle, S. P., Braun, P. G., Bos, R., Daykin, M., Camara, M., Williams, P., & Quax, W. J. (2006). Quorum quenching by an N-acyl-homoserine lactone acylase from *Pseudomonas aeruginosa* PAO1. *Infection and Immunity, 74*, 1673–1682.

Smadja, B., Latour, X., Faure, D., Chevalier, S., Dessaux, Y., & Orange, N. (2004b). Involvement of N-acylhomoserine lactones throughout plant infection by *Erwinia carotovora* subsp. *atroseptica* (*Pectobacterium atrosepticum*). *Molecular Plant-Microbe Interactions, 17*, 1269–78.

Smadja, B., Latour, X., Trigui, S., Burini, J. F., Chevalier, S., & Orange, N. (2004a). Thermodependence of growth and enzymatic activities implicated in pathogenicity of two *Erwinia carotovora* subspecies (*Pectobacterium* spp.). *Canadian Journal of Microbiology, 50*, 19–27.

Smid, E. J., Jansen, A. H. J., & Tuijn, C. J. (1993). Anaerobic nitrate respiration by *Erwinia carotovora* subsp. *atroseptica* during potato tuber invasion. *Applied and Environmental Microbiology, 59*, 3648–3653.

Smith, K. M., Bu, Y., & Suga, H. (2003a). Induction and inhibition of *Pseudomonas aeruginosa* quorum sensing by synthetic autoinducer analogs. *Chemistry & Biology, 10*, 81–89.

Smith, K. M., Bu, Y., & Suga, H. (2003b). Library screening for synthetic agonists and antagonists of a *Pseudomonas aeruginosa* autoinducer. *Chemistry & Biology, 10*, 563–571.

Sproer, C., Mendrock, U., Swiderski, J., Lang, E., & Stackebrandt, E. (1999). The phylogenetic position of *Serratia*, *Buttiauxella* and some other genera of the family *Enterobacteriaceae*. *International Journal of Systematic Bacteriology, 49*, 1433–1438.

Steidle, A., Sigl, K., Schuhegger, R., Ihring, A., Schmid, M., Gantner, S., Stoffels, M., Riedel, K., Givskov, M., Hartmann, A., Langebartels, C., & Eberl, L. (2001). Visualization of N-acylhomoserine lactone-mediated cell–cell communication between bacteria colonizing the tomato rhizosphere. *Applied and Environmental Microbiology, 67*, 5761–5770.

Taguchi, F., Ogawa, Y., Takeuchi, K., Suzuki, T., Toyoda, K., Shiraishi, T., & Ichinose, Y. (2006). Homologue of 3-oxoacyl-(acyl carrier protein) synthase III gene located in glycosylation island of *Pseudomonas syringae pv. tabaci* regulates virulence factors via N-acyl homoserine lactone and fatty acid synthesis. *Journal of Bacteriology, 188*, 8376–8384.

Teplitski, M., Chen, H., Rajamani, S., Gao, M., Merighi, M., Sayre, R. T., Robinson, J. B., Rolfe, B. G., & Bauer, W. D. (2004). *Chlamydomonas reinhardtii* secretes compounds that mimic bacterial signals and interfere with quorum sensing regulation in bacteria. *Plant Physiology, 134*, 1–10.

Teplitski, M., Robinson, J. B., & Bauer, W. D. (2000). Plants secrete substances that mimic bacterial N-acyl homoserine lactone signal activities and affect population density-dependent behaviors in associated bacteria. *Molecular Plant-Microbe Interactions, 13*, 637–648.

Toth, I. K., Newton, J. A., Hyman, L. J., Lees, A. K., Daydin, M., Ortori, C., Williams, P., & Fray, R. G. (2004). Potato plants genetically modified to produce N-acylhimoserine lactones increase susceptibility to soft rot *Erwiniae*. *Molecular Plant-Microbe Interactions, 17*, 880–887.

Tsuyumu, S. (1977). Inducer of pectic acid lyase in *Erwinia carotovora*. *Nature, 269*, 237–238.

Uroz, S., Chhabra, S. R., Càmara, M., Williams, P., Oger, P. M., & Dessaux, Y. (2005). N-acylhomoserine lactone quorum-sensing molecules are modified and degraded by *Rhodococcus erythropolis* W2 by both amidolytic and

novel oxidoreductase activities. *Microbiology, 151*, 3313–3322.
Uroz, S., Dangelo, C., Carlier, A., Faure, D., Petit, A., Oger, P., Sicot, C., & Dessaux, Y. (2003). Novel bacteria degrading N-acyl homoserine lactones and their use as quenchers of quorum-sensing regulated functions of plant pathogenic bacteria. *Microbiology, 149*, 1981–1989.
Uroz, S., Oger, P., Chhabra, S. R., Camara, M., Williams, P., & Dessaux, Y. (2006). N-acyl homoserine lactones are degraded via an amidolytic activity in *Comamonas* sp. strain D1. *Archives of Microbiology, 187*, 249–256.
Von Bodman, S. B., Bauer, W. D., & Coplin, D. L. (2003). Quorum sensing in plant-pathogenic bacteria. *Annual Review of Phytopathology, 41*, 455–482.
Wang, Y. J., & Leadbetter, J. R. (2005). Rapid acyl-homoserine lactone quorum signal biodegradation in diverse soils. *Applied and Environmental Microbiology, 71*, 1291–1299.
Wang, C., Zhang, H. B., Wang, L. H., & Zhang, L. H. (2006). Succinic semialdehyde couples stress response to quorum-sensing signal decay in *Agrobacterium tumefaciens*. *Molecular Microbiology, 62*, 45–56.
Waters, C. M., & Bassler, B. L. (2005). Quorum sensing: Cell-to-cell communication in bacteria. *Annual Review of Cell and Developmental Biology, 21*, 319–346.
Wei, Z. M., & Beer, S. V. (1995). *hrpL* activates *Erwinia amylovora hrp* gene transcription and is a member of the, E.C.F subfamily of sigma factors. *Journal of Bacteriology, 177*, 6201–6210.
Wei, Z. M., Laby, R. J., Zumoff, C. H., Bauer, D. W., He, S. Y., Collmer, A., & Beer, S. V. (1992). Harpin, elicitor of the hypersensitive response produced by the plant pathogen *Erwinia amylovora*. *Science, 257*, 85–88.
Whitehead, N. A., Barnard, A. M. L., Slater, H., SNJ, L., & Salmond, G. P. C. (2001). Quorum-sensing in Gram-negative bacteria. *FEMS Microbiology Review, 25*, 365–404 .
Whitehead, N. A., Byers, J. T., Commander, P., Corbett, M. J., Coulthurst, S. J., et al. (2002). The regulation of virulence in phytopathogenic *Erwinia* species: Quorum sensing, antibiotics and ecological considerations. *Antonie van Leeuwenhoek, 81*, 223–231.
Williams, N. (2002). Back to the public over the, G. M. crop battle. *Current Biology, 12*, R605–R606.
Williamson, L. L., Borlee, B. R., Schloss, P. D., Guan, C., Allen, H. K., & Handelsman, J. (2005). Intracellular screen to identify metagenomic clones that induce or inhibit a quorum-sensing biosensor. *Applied and Environmental Microbiology, 71*, 6335–6344.
Wisniewski, J. P., Frangne, N., Massonneau, A., & Dumas, C. (2002). Between myth and reality: Genetically modified maize, an example of a sizeable scientific controversy. *Biochimie, 84*, 1095–1103.
Xu, F., Byun, T., Deussen, H.-Y., & Duke, K. R. (2003). Degradation of N-acylhomoserine lactones, the bacterial quorum-sensing molecules, by acylase. *Journal of Biotechnology, 101*, 89–96.
Yang, C. H., Gavilanes-Ruiz, M., Okinaka, Y., Vedel, R., Berthuy, I., Boccara, M., Chen, J. W., Perna, N. T., & Keen, N. T. (2002). *hrp* genes of *Erwinia chrysanthemi* 3937 are important virulence factors. *Molecular Plant-Microbe Interactions, 15*, 472–480.
Yang, F., Wang, L. H., Wang, J., Dong, Y. H., Hu, J. Y., & Zhang, L. H. (2005). Quorum quenching enzyme activity is widely conserved in the sera of mammalian species. *FEBS Letters, 579*, 3713–3717.
Yap, M. N., Rojas, C. M., Yang, C. H., & Charkowski, A. O. (2006). Harpin mediates cell aggregation in *Erwinia chrysanthemi* 3937. *Journal of Bacteriology, 188*, 2280–2284.
Yates, E. A., Philipp, B., Buckley, C., Atkinson, S., Chhabra, S. R., Sockett, R. E., Goldner, M., Dessaux, Y., Camara, M., Smith, H., & Williams, P. (2002). N-acyl homoserine lactones undergo lactonolysis in a pH-, temperature-, and acyl chain length-dependent manner during growth of *Yersinia pseudotuberculosis* and *Pseudomonas aeruginosa*. *Infection and Immunity, 70*, 5635–5646.
Zhang, L. H. (2003). Quorum quenching and proactive host defense. *Trends in Plant Science, 8*, 238–244.
Zhang, H. B., Wang, L. H., & Zhang, L. H. (2002). Genetic control of quorum-sensing signal turnover in *Agrobacterium tumefaciens*. *Proceedings of the National Academy of Sciences of the USA, 99*, 4638–4643.
Zhang, H. B., Wang, C., & Zhang, L. H. (2004). The quormone degradation system of *Agrobacterium tumefaciens* is regulated by starvation signal and stress alarmone (p)ppGpp. *Molecular Microbiology, 52*, 1389–1401.

MIX
Papier aus verantwortungsvollen Quellen
Paper from responsible sources
FSC® C105338

If you have any concerns about our products,
you can contact us on
ProductSafety@springernature.com

In case Publisher is established outside the EU,
the EU authorized representative is:
Springer Nature Customer Service Center GmbH
Europaplatz 3, 69115 Heidelberg, Germany

Printed by Libri Plureos GmbH
in Hamburg, Germany